Applied Design Research

A Mosaic of 22 Examples, Experiences and Interpretations Focussing on Bridging the Gap between Practice and Academics

Edited by Peter Joore, Guido Stompff, Jeroen van den Eijnde

First edition published 2022
by CRC Press
6000 Broken Sound Parkway NW, Suite 300, Boca Raton, FL 33487–2742
and by CRC Press
4 Park Square, Milton Park, Abingdon, Oxon, OX14 4RN

CRC Press is an imprint of Taylor & Francis Group, LLC

Library of Congress Cataloging-in-Publication Data

Names: Joore, Peter, 1967- editor. | Stompff, Guido, editor. | Eijnde, Jeroen van den, editor.

Title: Applied design research: a mosaic of 22 examples, experiences and interpretations focussing on bridging the gap between practice and academics / edited by Peter Joore, Guido Stompff, Jeroen van den Eijnde.

Description: First edition. | Boca Raton, FL: CRC Press, 2022. | Essays translated from Dutch. | Includes bibliographical references.

Identifiers:
LCCN 2021059382
ISBN 9781032209197 (hbk)
ISBN 9781032209173 (pbk)
ISBN 9781003265924 (ebk)

Subjects: LCSH: Human engineering. | Industrial design. | Design--Social aspects. | Design--Research.
Classification: LCC TA166 .A673 2022 | DDC 620.8/2--dc23/eng/20220110
LC record available at https://lccn.loc.gov/2021059382

DOI: 10.1201/9781003265924

This publication is a result of the NADR2 project, executed by the Network Applied Design Research. This project was co-funded by Taskforce SIA, part of the Netherlands Organisation for Scientific Research (NWO).

Contributors: Peter Joore, Guido Stompff, Jeroen van den Eijnde, Daan Andriessen, Karin van Beurden, Rens Brankaert, Anke Coumans, Tessa Cramer, Wander Eikelboom, Tomasz Jaskiewicz, Christine de Lille, Remko van der Lugt, Masi Mohammadi, Sebastian Olma, Anja Overdiek, Eke Rebergen, Perica Savanović, Wina Smeenk, Aletta Smits, Peter Troxler, Koen van Turnhout, Job van 't Veer, Eveline Wouters, Marieke Zielhuis, Antien Zuidberg.

Design and layout: Studio RATATA.nl
Illustrations: Kalle Wolters
Translation: Proactive Translations

Publisher's note: This book has been prepared from camera-ready copy provided by the authors.

Content

Preface   6
Karin van Beurden

About the editors   8
Peter Joore, Guido Stompff, Jeroen van den Eijnde

Applied Design Research   10
Peter Joore, Guido Stompff, Jeroen van den Eijnde

Part 1: Eyes on the future

Research into research   25
Daan Andriessen

Radio Dabanga   33
Koen van Turnhout & Aletta Smits

Learning from prototypes   43
Tomasz Jaskiewicz

Design thinking for professionals   53
Guido Stompff

Dance? Dance!   63
Peter Troxler

Part 2: The urge to improve the world

**Idealistic visions of the future
or realistic solutions?**   75
Peter Joore

Designing the future   85
Tessa Cramer

The artistic attitude in a social context   95
Anke Coumans

Looking for trouble   105
Eke Rebergen, Sebastian Olma, Wander Eikelboom

Discomfort as a starting point   115
Marieke Zielhuis

4

Part 3: Design and research with others

Systemic co-design   127
Remko van der Lugt

Inclusive designs in healthcare   137
Rens Brankaert

Societal impact design   147
Wina Smeenk

Designing our society together   157
Christine De Lille

Integral development of the built environment   167
Perica Savanović

Part 4: Building bridges between disciplines

Smart transitions with design   179
Anja Overdiek

A new mindset in research   189
Eveline Wouters

Focus on the practical question   197
Job van 't Veer

Shaping an empathic living environment   207
Masi Mohammadi

Seducing the conshuman   217
Antien Zuidberg

Part 5: The task for applied design research

Something old, something new   229
Karin van Beurden

A letter from the future   241
Jeroen van den Eijnde

In conclusion

Epilogue   252
Peter Joore, Guido Stompff, Jeroen van den Eijnde

Preface

Karin van Beurden

Design research, a form of research that is highly interwoven with design, is a relatively new discipline in the world of research. The first "design generation" in the 1960s tended towards a rational and systematic development of the field.[1] Inspired by visionary designers like Victor Papanek,[2] in time, more social-oriented approaches came up. Being a design student, I followed Papanek's design class in 1979 at Kansas City Art Institute, a couple of years after he published his now-famous book *Design for the Real World: Human Ecology and Social Change*. I have always remembered Papanek's inspiring lessons while working as a product designer and design researcher myself.

From 2001 onward, at universities of applied sciences in the Netherlands, research groups were founded that started working on an applied form of design research. Each of these research groups developed its own specific approach. One focuses on healthcare, others on circular innovations or on innovative networks. Even though they differ significantly in size, background, and focus, these applied design research groups do share a common language and approach. When professor Daan van Eijk asked me, in 2012, whether I wanted to represent the applied research sector in CLICK | Design,[3] I agreed to do so after consulting a handful of design professors from other universities. Although I didn't have an exact picture of who or what I was representing at that time, in retrospect this was the moment when the seed for the Network Applied Design Research (NADR) was planted.

Mutual contacts and intense discussions on approach, used methods and best practices, led to the establishment of the Network Applied Design Research in 2016. A network of researchers who all focus on applied design research,

albeit within very different domains. The network has since grown into an intensively collaborating group of highly committed professors and researchers, focusing on applied design research. Through the Network Applied Design Research, we aim to encourage collaboration between research groups and enable discussion about how design and research can mutually enhance each other.

In addition to promoting and identifying high-quality design research, another vital activity of the network is increasing the visibility of the field. This is also the purpose of this book: to show the diversity of applied design research, in all its aspects. At the same time, the book is also a form of self-reflection, in which 25 passionate researchers reflect on what they do, their approach, and what they consider impor- tant. It shows the reader what the authors have in common: their focus on the future, the drive to change, and the inten- sive collaboration with users and other disciplines.

The initiative for this publication lies with Peter Joore. After years of fascinating discussions, many open conversations full of new insights, and moments of recognition ("that's exactly what I mean and what makes our profession so disdistinctive") it's thanks to him that our experiences have been brought together in this publication, allowing them to find their way to a wider audience. The editorial team, con- sisting of Peter Joore and co-editors Jeroen van den Eijnde and Guido Stompff, has successfully merged the very diverse contributions into an appealing and accessible book. As the chairperson of the Network Applied Design Research, I am particularly proud that this book was created with the input of so many. And I am sure it will be an inspiration to anyone interested in applied design research!

Karin van Beurden
Chairperson of the Network Applied Design Research
Professor of Product Design at Saxion University of
Applied Sciences

1.   Harriet Atkinson,
Maya Rae Oppenheimer,
"Design Research – History,
Theory, Practice: Histories
for Future-Focused
Thinking," *Proceedings of
Design Research Society
50th Anniversary Conference*
(Brighton, June 2016).

2.   Victor Papanek, *Design
for the Real World. Human
Ecology and Social Change*
(St Albans: Paladin, 1974).

3.   CLICK | Design was the
forerunner of CLICKNL, the
innovation network of the
Dutch Creative Industry.

About the editors

Peter Joore, NHL Stenden University of Applied Sciences

Dr.ir. Peter Joore focuses on design processes in which different actors, across sectoral boundaries, work together to solve complex societal issues in a living lab environment. He was trained as an industrial designer at TU Delft. He worked as a designer at several companies, among others working on a redesign of the Fokker 50 aircraft interior, the development of signage for Hong Kong's Mass Transit Railway Cooperation and the design of check-in systems for Moscow's Domodedovo Airport. In 2008, he switched to higher education, working as a professor of Open Innovation at NHL Stenden University of Applied Sciences in Leeuwarden.

Guido Stompff, Inholland University of Applied Sciences

Dr.ir. Guido Stompff has been a Professor of Design Thinking at the Creative Business research group of the Inholland University of Applied Sciences since 2019. After his training as an industrial designer (TU Delft), he worked for over 25 years as a designer, covering the full scope of the field, including product design, UX design, communication design, packaging design, branding, and even art. In 2011, he obtained his PhD in the facilitation of innovation in multidisciplinary teams. His book *Design thinking, radicaal veranderen in kleine stappen* was published in 2018. The book was voted Dutch management book of the year.

Jeroen van den Eijnde, Artez University of the Arts

Dr. Jeroen van den Eijnde studied product design at the Arnhem Art Academy and art history at Leiden University. He obtained his PhD with a study into the theory and ideology in Dutch formgiving education. Since 2016, he has worked as a Professor of Tactical Design at ArtEZ University of the Arts. Van den Eijnde was co-founder and board member of the Design Platform Arnhem. As a consultant, he worked for the Fonds Beeldende Kunst, Vormgeving en Bouwkunst (the Fine Art, Design and Architecture Foundation) and the Raad voor Cultuur (Culture Council). He is member of the program council for CLICKNL, the innovation network of the creative industry's top sector.

Network Applied Design Research

All editors and authors in this book are involved in the Network Applied Design Research, NADR. NADR is a joint initiative of design researchers affiliated with different universities of applied sciences. They have joined forces and work together to ensure the quality and visibility of applied design research. The NADR partners apply design research within a broad range of industries and work among others within the healthcare, food and agriculture sector, the built environment, and on the development of circular products and services. For more information, see www.nadr.nl.

Applied Design Research

Peter Joore, Guido Stompff, Jeroen van den Eijnde

Design and research: two areas of expertise, each with its own traditions, methods, standards, and practices. Two worlds that are still quite rigorously separated: researchers research what is there, designers imagine what could be possible. Design research is trying to bridge the gap by integrating design and research to develop new knowledge. However, building bridges between two worlds is not easy. The search for what design research is, how to perform this type of research, and which standards must be met resulted in a proliferation of terms, such as *research through design,*[1] *speculative design,*[2] or *design research through practice.*[3] The lively discussions on definitions, concepts, and methods show that the field is developing and there is growing interest from far beyond the design world.

For this book, we chose the term *Applied Design Research.* "We" are the Network Applied Design Research: a learning community of professors and researchers at various universities of applied sciences. Within this community, we share our experiences with the many forms of applied design research. We consciously choose the word "applied" because we like to emphasize the practical application of design research. So what do we mean by applied design research? And do we even have a shared understanding about the concept?

By asking the members of the network: "What is your defi-nition of applied design research?", we created a unique spectrum of different perspectives. The resulting articles give a glimpse into the kitchen of twenty-five professors and researchers who apply this inspiring approach to product development, architecture, the arts, healthcare, food, and the social sector. It leads to a remarkably transdisciplinary research field. When reading and discussing the articles, we found a robust pattern of characteristics. These all manifest to a greater or lesser extent in each individual contribu-tion, including a future-orientation, the desire to improve the world and the ambition to involve others in the design process.

The book, organized into five parts, has not become a recipe book but rather a mosaic of articles, each offering a different interpretation, different illustrative examples, and different methods. The characteristic of a mosaic is that each piece contributes to the whole, but none contains all the infor-mation. Together, they offer an excellent picture of applied design research, its use, and what you can expect from it. We hope that the book, with its many examples, can inspire (novice) researchers to start applying this inspiring approach. And that more experienced design researchers recognize themselves in this book and feel challenged by it.

1. William W. Gaver, "What Should We Expect From Research Through Design?," in *Proceedings of the 2012 ACM Annual Conference on Human Factors in Computing Systems* (May 2012): 937–946, https://doi.org/10.1145/2207676.2208538.

2. Anthony Dunne and FionaRaby, *Speculative Everything; Design, Fiction, and Social Dreaming* (Cambridge, MA: MIT Press: 2013).

3. Ilpo Koskinen, John Zimmerman, Thomas Binder, Johan Redström and Stephan Wensveen, *Design Research Through Practice: From The Lab, Field, and Showroom* (Amsterdam: Elsevier, 2011).

11

Part 1: Eyes on the future

The researcher is focused on understanding the world as we know it; the designer is focused on developing alternative futures. Applied design research combines both and deals with what is desired and thus tells us the current problems. But because the future does not yet exist, it is also difficult to discuss it. This appeals to the unique quality of the designer: being able to visualize things that cannot be addressed, in the form of visualizations, objects, or interventions.

The envisioning of future realities generates new knowledge, "knowing." Applied design research is a unique form of science that tries to initiate intentional changes, to direct the flow of events towards a more desired future. However, this calls for difficult methodological and epistemological questions. The first part of this book contains articles by lecturers that address the methodological challenge of outlining a theoretical context for applied design research.

Daan Andriessen, in *"Research Into Research,"* describes the development of design science research, and indicates that in that field, the design of the solution itself, until recently, was mainly seen as a "creative leap" that was less important to the final result of the research. Over time, he has become increasingly interested in design research that involves developing real solutions for pressing problems. Andriessen wonders how design can be used more effectively as a knowledge-generating activity.

He was inspired by, among other things, the collaboration with *Koen van Turnhout.* In *"Radio Dabanga,"* Koen and Aletta Smits, through an illustrative case study, offer insights into different types of knowledge that are characteristic of applied design research. The two distinguish between knowledge about the current situation, the desired futures, and effective solutions to get there. In other words, knowledge of how it |is|, how it |can| be, and how it |will| be when we apply effective solutions. Each of these types of knowledge has different quality criteria.

This taxonomy echoes in the results of *Tomasz Jaskiewicz's* research into practical knowledge acquired by designers. In *"Learning From Prototypes."* he distinguishes between insights, ideas and know-how. These terms describe, respectively, what designers consider to be truth, what are speculative assumptions, and what needs to be done to achieve something.

This taxonomy is also recognizable in *"Design Thinking for Professionals,"* by *Guido Stompff,* in which he provides a process description for design research for professionals who do not have a design background. Not only designers design; every professional designs from time to time. In an iterative process of interpretation (of the existing situation), envisioning (the desired situation) and design (how to get there), a new framework and new knowledge are being developed along the way. In this process, envisioning can radically change the problem definition, just as design proposals can question envisioned goals.

In *"Dance? Dance!"* author *Peter Troxler* describes the know-how of designers to use artifacts as boundary objects to bring together different parties and stakeholders, offering a common language. These *boundary objects* are recognizable objects that can mean very different things for the various parties involved. When language is lacking, such objects enable people to gain joint insights and come up with solutions in a *ballet of disciplines.*

13

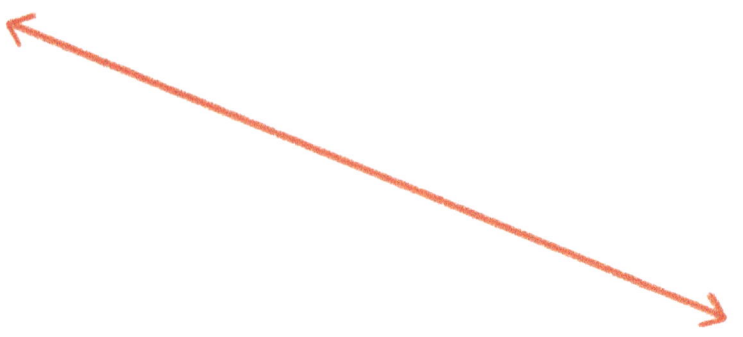

Part 2: The urge to improve the world

In the second part, we highlight another characteristic of applied design research. In the sciences, there is broad consensus to strive for value-free research. This endeavor is commendable to understand the current world, but less valuable if you want to change it (a bit). Value-free research can explain the causes and consequences but cannot indicate what is desired or how to get there, let alone help develop new perspectives.

In every article in this book, the intention of the researchers to improve the world can be felt. Applied design research is anything but value-free. *Peter Joore* succinctly explains that in his article *"Idealistic Visions of the Future or Realistic Solutions?":* "It is ultimately about developing a better and more beautiful world, one in which idealistic visions of the future are translated into realistic solutions in the here and now." That makes applied design research normative, because it strives for the "good," such as a more sustainable world, a more inclusive society or better health care. But it also raises delicate questions, such as what is "better"? For whom? And why is it better? Where the researcher is focused on indicating what is true, and thus on making a *truth* judgment, the designer is focused on creating value and on making a *value* judgment. This enables research and design to complement each other.

The question 'what is desired' is also difficult to answer because it is hard to envision what is possible in the future. For example, in *"Designing the Future,"* author *Tessa Cramer* describes that creating future solutions "requires fresh perspectives that old systems cannot think of" and that students need to hack into those systems "in a friendly way." In doing so, she chooses a speculative approach to make people think, with her goal being to teach people to deal with uncertainty. She calls for a forward-looking, design-oriented approach because creatives, artists, and designers can contribute to thought processes around societal issues.

This is in line with *Anke Coumans'* perspective, who in *"The Artistic Attitude in a Social Context"* describes how students work to portray elderly people with dementia, not so much to solve problems quickly, but to empathize with others. Instead of focusing on the development of solutions, this perspective is mainly about understanding and explaining a particular situation. She aims to design environments in a way that enables other professionals, for example in the healthcare industry, to start acting as designers.

Eke Rebergen, Sebastian Olma and *Wander Eikelboom* expand on this in *"Looking for Trouble."* Instead of being troubleshooters, designers must "question the socio-cultural and ethical consequences of technological developments and use design to demonstrate that the world can be different." They want designers to have a critical attitude and suggest a form of speculative design "to raise, gather and own problems." This can be annoying or uncomfortable.

In her article, *"Discomfort as a Starting Point,"* Marieke Zielhuis chooses a different perspective. She describes her "discomfort" when she discovers that designers are seen primarily as problem solvers, providing solutions in collaborations. This ignores the needs of designers to develop in their new role. That is why she does not study what designers should do, but what they need to grow. In her research, she focuses on the question, "How can design research contribute to the design practice?"

15

Part 3: Design and research with others

In all articles, a human-centered perspective manifests, sometimes implicitly but usually explicitly. A focus on humans who do not always act rationally, make intuitive choices and behave differently depending on the situation. Words like "empathy" are among the most commonly used in this book. However, the unruly complexity of *applied* design research lies in the fact that each design study involves many people. They all have different interests, different opinions, and different reactions, showing con-trasting perspectives that are nearly impossible to bridge. A reductionist, analytical perspective fails to understand and change such situations.

Applied design research brings a fresh approach, trying to transcend these contradictions. Devising and testing pro-totypes is intended not so much to discuss the contrasting perspectives, but to create a design that is tested in practice: what works and what doesn't? Does it help people to achieve a goal? How is it experienced? What surprises? The feedback provides knowledge for the next iteration of design. Applied design research embraces subjectivity, but refrains from "anything goes" relativism by putting practical consequences central.

Some researchers in this book go beyond testing in order to involve people. In *"Systemic co-design,"* Remko van der Lugt discusses his extensive experience in facilitating co-design processes. No longer is design framed as designing *for* others, but as designing *with* others. His research centers around how all participants can develop their design capabilities so that they can contribute fully to the design process. And how (professional) designers can take on a role that facilitates that joint process, while also being able to express their own creativity. This is an important question, as it would be such a waste if experienced designers cannot use their knowledge and skills.

Rens Brankaert is active in the healthcare industry and works with teams of healthcare professionals, design professionals, and design researchers. In *"Inclusive Design in Healthcare,"* he describes how design research can be used to develop "warm technology": inclusive technology that focuses on what someone is still able to do. The challenge he sees, in line with what Guido Stompff writes in another article, is to provide healthcare professionals with sufficient design skills to start design research in their own practical environment.

In a way, this challenge has already been tackled by *Wina Smeenk*. In *"Societal Impact Design,"* she describes the dementia simulator. Through a visit to this simulator, healthy people (healthcare professionals and informal caretakers) experience what it is like to live with dementia, enabling them to better empathize with the people involved. And this empathy will make them act differently. She emphasizes that there are no simple solutions for many of the issues; they require a holistic and empathetic look, considering the system as a whole.

Co-design is not only applicable to healthcare, as is shown in *Christine de Lille's* article. In *"Designing Our Society Together,"* she describes co-design between actors in the retail industry. In the Future Proof Retail project, a project also involving Anja Overdiek, the team developed 22 (!) living labs where industry associations, local retailers, and their employees worked together on solutions that enabled them to become future-proof. The labs shared their experiences, and the results enabled the authorities to support the retail industry on a larger scale. Based on their experiences, they also published a how-to manual for innovation in labs.

17

In *"Integral Development of the Built Environment,"* author *Perica Savanović* discusses co-design in the context of building, although he refrains from using the word co-design. He focuses his lens on collaboration between construction experts, residents, and policymakers. He observes that the current construction practice hinders innovation because all design requirements have been specified in great detail beforehand, limiting the solution space. A shared design process enables outcomes to transcend the requirements, as all stakeholders jointly explore the design options before making final decisions.

Part 4: Building bridges between disciplines

It is remarkable that applied design research is used in all kinds of contexts. The articles focus on the healthcare industry, the construction industry, the social sector, retail, education, and the public domain. Design research is apparently relevant to all those different worlds, each with its own knowledge, habits, practices, and stakeholders.

For example, in *"Smart Transitions With Design,"* Anja Overdiek describes her own background as a sociologist and psychologist. As an expert in scientific theorization, she had turned her back on science after graduating: "To me, it was too much like working in an ivory tower." She describes how the discovery of applied research lured her back to the world of research because this approach enables her to keep grounded in practices and effectively connect problems, distinct frames of people with future possibilities.

Eveline Wouters also sees herself as the odd one out in her article *"A New Mindset in Research."* She is not a "real" designer, but has a medical background. She describes how, over the years, she has come into contact with a broadening arsenal of research methods, with design research taking

18

a special place. She considers applied design research to be more than a research method; it is a *mindset* in which the complete involvement of the end-users in the design of a product, service, or organizational change is of vital importance.

In *"Focus on the Practical Question,"* Job van 't Veer focused on that same mindset when discussing his healthcare and welfare students. They are attracted to design methods that are aimed at empathy and co-creation with the target group. He found that, although there are plenty of books about the many forms of design research, none of them were aligned to healthcare education. This led to a textbook focused on design-oriented work for (higher professional) healthcare and welfare education, that he wrote with Eveline Wouters and Remko van der Lugt, both of whom contributed to this book.

In *"Shaping an Empathic Learning Environment,"* Masi Mohammadi describes the fertile crossroads of technology, healthcare, and construction, emphasizing the latter. These are industries where significant changes are already taking place, especially when you start working at the cutting edge of these industries. She describes, among other things, the development of the "Empathic Home," which was developed in close collaboration with a wide range of companies, housing corporations, and healthcare organizations.

Antien Zuidberg describes another area of application in *"Seducing the Conshuman."* She explains how applied design research is used in the food and agriculture industry. As a food technologist, she worked in the food industry for years. She worked, among other things, on the application of proteins in food products. She observed that although this industry had a rapid technological development, a more significant transition was needed to become sustainable and healthier. She is now using her Food Innovation Model to entice the "consumer being" to take steps in the right direction.

19

Part 5: The task for applied design research

Applied design research is being applied more and more, but where is it heading? In *"Something Old, Something New,"* NADR chair *Karin van Beurden* looks back on her 40 years of experience and names the task we face, a challenge that Victor Papanek [4] already posed to the design world half a century ago. He called for contributions to a better world and to stop designing poor products that sell well. Karin describes her development through a multitude of examples. On the one hand, much has changed in the design world, but on the other hand, the challenge has remained the same, namely, to develop solutions that are – in Papanek's words – "responsive to the true needs of mankind." And apparently, that is much more difficult than people think.

There is a light at the end of the tunnel. The contributions show that applied design research is a form of research that distinguishes itself fundamentally from other research. It is generative, imaginative, future-oriented, and challenges people to alter the course of unfolding events into a more preferred future. It enables designers, users, residents, experts, and other stakeholders to contribute to the challenges we face. Applied design research has no issues with using the word "better," no matter how tricky the discussion gets. And it can be applied in a variety of sectors, although much development is still needed. At the same time, applied design research is critical of itself and warns against technocratic solutionism. With this, applied design research can contribute to shape our collective futures for the better. In the final contribution, *Jeroen van den Eijnde* looks back on this with his *"Letter From the Future. An Attempt at Good Ancestry."*

4. Victor Papanek, *Design for the Real World; Human Ecology and Social Change* (St Albans: Paladin, 1974).

In conclusion

Imagining plays a central role in applied design research. Imagining is expressed by inspiring visualizations, tangible objects, or meaningful experiences. As the English say, one should "practice what you preach," which is why we asked illustrator Kalle Wolters to express each article by means of one illustration. He surprised us with his interpretations. Just as a good poem can convey an emotion that we cannot describe adequately, he managed to distill something important from each article that threatened to "drown" in the jumble of words. Imagining requires a certain level of courage because intuition, skill, and empathy are needed to portray what is (still) difficult to express in words.

PART 1: EYES ON THE FUTURE

"Discovering laws involves drafting them. Recognizing patterns is very much a matter of inventing and imposing them. Comprehension and creation go on together."

~ **Nelson Goodman**

Research into research

Design processes as participatory knowledge production

Daan Andriessen

In the autumn of 1999, I met my intended supervisor Mathieu Weggeman in a grubby roadside restaurant somewhere in the Dutch province of Brabant. I planned to obtain my PhD by answering the question of how companies can measure what the knowledge in their company is worth. My problem was that I had not yet started the research, even though I had already designed the solution and even partly tested it. 'Well,' Mathieu said, 'you should read Joan van Aken's 1994 and 1996 articles on design science research. They discuss a methodology for doing research aimed at designing and testing methods such as yours.' That was my first introduction to design research.

For eight years now, we have been conducting research at the Utrecht University of Applied Sciences on practical research methodology. Design research – in all its variants – is an important trend within this methodology. So we are, in fact, researching research. Our focus is on the research competence of students, lecturers, and researchers at the universities of applied sciences. We help education programs and lecturers to get a good impression of the students' research competence through their curricula and graduation efforts. We help researchers master practice-oriented research that has an impact on practice, and we develop tools for better research.

Our research focuses on five issues: 1) How are research competences expressed in the context of education, research, and professional practice, and what is the relationship with the other capabilities of professionals? 2) Which forms of methodical thoroughness are appropriate for the use of research competence? 3) How can collaboration between disciplines and those involved in research be promoted? 4) How can the effect of research competence be increased? And 5) How can the mind shift and organizational transformation that are involved in research competence be stimulated?

One of the ongoing studies in the research group is the research of PhD student Marieke Zielhuis (see elsewhere in this collection). She has a background as a designer (TU Delft) and looks at how design research can yield more for designers in a practical environment. In many cases, the focus of design research is on developing knowledge and products to solve specific problems. Such research can also provide valuable results that can help you design better. However, such knowledge does not yet sufficiently reach the design practice. How can that be improved?

Design science research as social science designing

I applied Joan van Aken's approach in my thesis and obtained a PhD from Nyenrode University in 2003 with a study in which I had designed a valuation method and tested it at six companies. Not long after, I met Joan van Aken in person. In 2006 we founded the Design Science Research Group (DSRG),[1] a community of practice for researchers with a common interest in the methodology of this kind of research.

Soon, I discovered that the design science research trend, with authors such as Van Aken and Romme, is dominated by management sciences and is relatively separate from other design trends such as technology, IT and industrial design.

Design science research is, as it were, the social scientific variant of design research. The DSRG, therefore, mainly attracts other social scientific researchers, for example, from the educational field. Design science research has many affinities with educational design research, with authors such as Van den Akker and McKenney.

This social scientific variant designs social artifacts, not physical ones: ways of action that are contained in methods and concrete interventions. In other words, we design 'things you can do' such as a project management method ,[2] a method for setting up communities of practice ,[3] or a method for setting up hybrid learning environments .[4] These are solutions to practical problems, which are validated in a social scientific way.

For me, this also explains the increasing popularity of this type of research, especially at universities of applied sciences. The practical researchers at the universities of applied sciences want to contribute to a better practical environment by providing solutions .[5] Design science research is focused on this. At the same time, it is an approach that focuses on the scientific validation of that solution. I meet many lecturers at universities of applied sciences who find this combination attractive and perceive it as a way to promote and at the same time contribute to the practical environment.

Another characteristic of design science research is that the design process is a means of achieving the solution, but that the designing itself is hardly used as a means of gaining insight. Design science research is therefore not research through design .[6] In fact, in the early years of the DSRG, the designing of the solution itself was mainly seen as a 'creative leap' that was less important for the end result. Instead, the focus was on achieving a proper definition of the problem and on validating the solution. However, in recent years, we have gained more insight into the design process as a separate research activity. The participation of researchers who have been trained as designers, such as professor Koen van Turnhout, has contributed greatly to this.

1.   www.dsrg.nl

2.   Nicoline Mulder, *Value-based Project Management. Een Aanpak voor Chaordische Projecten vanuit het Perspectief van het Complexiteitsdenken* (PhD Thesis, TU Eindhoven, 2012), https://doi.org/10.6100/IR740171.

3.   Donald Ropes, *Organizing Professional Communities of Practice* (Amsterdam: University of Amsterdam Press, 2010).

4.   Petra H.M. Cremers, *Designing Hybrid Learning Configurations; At the Interface Between School and Workplace*, PhD Thesis (Wageningen University, 10 February 2016).

5.   Jos De Jonge, *Praktijkgericht Onderzoek bij Lectoraten van Hogescholen* (The Hague: Rathenau Instituut, 2016).

6.   Pieter Jan Stappers and Elisa Giaccardi, "Research Through Design," in *The Encyclopedia of Human-Computer Interaction, 2nd edition, eds.* Mads Soegaard and Rikke Friis-Dam (Aarhus, Denmark: 2017): 1–94.

7. Daan Andriessen, "Kennisstroom and Praktijkstroom," in *Handboek Ontwerpgericht Wetenschappelijk Onderzoek; Wetenschap met Effect*, Joan Ernst Van Aken and Daan Andriessen, eds., (Den Haag: Boom Lemma, 2011): 79–93.

8. Joan Ernst Van Aken and Daan Andriessen, eds., *Handboek Ontwerpgericht Wetenschappelijk Onderzoek; Wetenschap Met Effect* (The Hague: Boom Lemma Publishers, 2011).

9. www.musework.nl

The third characteristic of design science research is that the studies are mostly 'small n' studies. It is about finding valid solutions to problems that are relatively unique and complex. For example, in the medical sciences (actually also a form of design research, since this also deals with designing and testing solutions to problems), large numbers of patients are often involved. Testing is done on a large scale in double-blind experiments, and statistics show whether the treatment works on average. On the other hand, design science research usually deals with small numbers. As a researcher, you will be lucky to find six organizations, for example, that deal with a particular problem for which you can develop a solution. Design science research, therefore, often means case study research with a solid qualitative character.

The starting point of design science research is that in such complex problems, every situation is unique. In each case, the solution you develop (the generic solution) must be made suitable for the local context (the specific solution). For example, one of the cases in my thesis was a small consultancy firm that told me in advance: 'Your method is interesting, but the lead time is three months. That is far too long for us. Can it be done in a day?' As it happens, it was possible.

Design science research has in common with other forms of design research that the researcher moves back and forth between the generic design and the concrete applications. In the DSRG, we started calling this 'moving between two streams': the knowledge stream and the practice stream.[7] This process is illustrated in Figure 1. We have now set up a third process in between, where the design activities take place. This shows the increasing attention within the DSRG for the design itself.

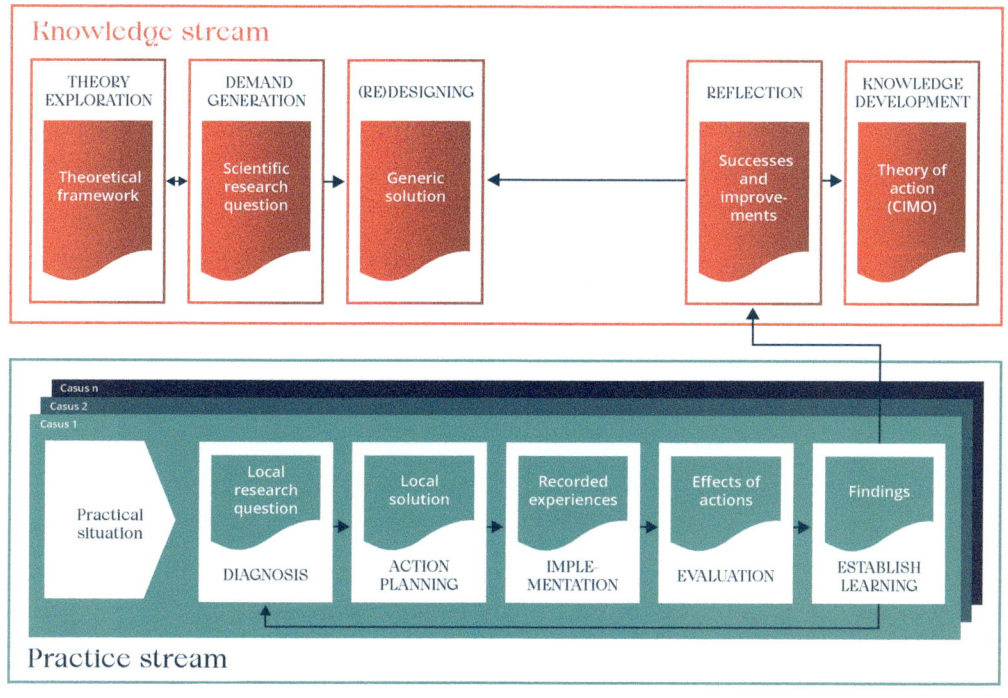

Figure 1
Knowledge stream and practice stream in design science research. [7]

More makership, more participation, and more recognition

One of the objectives of the DSRG is to develop the methodology of this type of research further. One of the first results was our 2011 *Handboek Ontwerpgericht Wetenschappelijk Onderzoek* (Handbook on Design-Driven Scientific Research).[8] In this manual, we elaborate on the above-mentioned characteristics and explain how you can solve bottlenecks in the approach. The methodological challenge we are now working on is to give the design process a more prominent place as an integral part of the methodology. This means more attention to design methods and to designing as a knowledge-generating activity.

One of my great sources of inspiration in this is the work of professor Bart van Rosmalen of the Utrecht School of the Arts. He has developed the concept of 'musework',[9] in which the researcher draws inspiration from the nine muses of Greek mythology. For example, Terpsichore, 'she who likes to dance'. How can dance, physical motion, and performance, be of service in practical research? Or Calliope, the muse of poetry, holding a stylus. How can poetry contribute to the development and transfer of new insights? A characteristic feature of all the muses is that they sing the praises of reality by making something. This makership can become even more central in design science research.

A second methodical challenge is to make the process of design science research more participatory. In the beginning, we tended to act as experts, coming in to research a problem, develop a solution from behind a desk, and come back to test the solution. I have also called design science research 'research by consulting', in which the role of consultant was more that of an expert than a process supervisor.

In recent years, we have become interested in getting stakeholders to participate in the research, and we have sought contact with action researchers. One of the results is a course called Action Research meets Design Research, which is hosted by four universities of applied sciences and the Vrije Universiteit Amsterdam. In this course, we examine the similarities and differences between design research and action research.

Finally, the challenge remains to get design science research acknowledged as a scientific approach in the social sciences. From the beginning, the social sciences have been very divided on the question of what is good scientific research, with many camps and paradigms. In this 'method struggle', the post-positivists have long had the upper hand. As a result, social scientific research was only taken seriously if it included statistics.

In recent years, there has been growing interest in other forms of research, partly due to a greater need for research that contributes to a better society. However, PhD students from the universities of applied sciences still find that their intended supervisor wants them to perform mainly descriptive and explanatory research; they do not accept design science research as an acceptable approach for a thesis.

Thanks to NADR, my colleagues at the Utrecht University of Applied Sciences, such as Remko van der Lugt and Koen van Turnhout, and Marieke Zielhuis' PhD research at TU Delft, I am increasingly involved in the world of design scientists. And I love that the focus on describing, explaining and statistics is almost non-existent in that world. I like that the focus is on developing solutions to pressing problems. Although that world has its own peculiarities ('Oh, you are not trained as a designer...I see...'), it is an exciting and inspiring environment in which I still want to learn a lot about better designing, validating, and further developing solutions for practical issues.

Daan Andriessen

Utrecht University of Applied Sciences

Dr. Daan Andriessen is Professor of Research Competence at the Utrecht University of Applied Sciences. He worked as an organizational consultant for KPMG during the first twelve years of his career. After obtaining his PhD from Nyenrode University in 2003, he started working at universities of applied sciences, first as a professor of Knowledge Management at Inholland and since 2013 as a professor at the Utrecht University of Applied Sciences. In his work, Daan tries to make research more relevant for practice, and he wants to use the complexity of the practical environment to enrich the world of research. He is interested in design research, action research, and in forms of research that apply the qualities of the arts.

31

Radio Dabanga

Applied design research in human experience & media design

Koen van Turnhout & Aletta Smits

Radio Dabanga is a radio station broadcasting for Sudan. However, their newsroom is located in Amsterdam because of Sudan's long history of repression of the free press. That means that two journalists are tasked with covering the journalistic needs of a population of 43 million citizens over 4000 miles away. How is that even possible? What are the information gathering and verification practices of these journalists? Can we design powerful tools that make this Herculean challenge even remotely manageable?

The research group Human Experience and Media Design (HEMD) aims to improve the user experience of digital media. Increasingly, user experience (UX) professionals working in design agencies, technology, or service companies have the opportunity to use data and AI as a design material to conceive new meaningful digital media products and to improve the quality of the current generation human-media interactions. Our focus is to support professionals in this challenge with inspirational examples, practical tools, methods and models. In practice, most of our projects form a rich amalgam of multiple horizons, stakeholders and complementary knowledge products. As the Radio Dabanga project is a vivid example of all these characteristics, we will use it throughout this paper to illustrate our approach to applied design research.

1. Koen van Turnhout, Arthur Bennis, Sabine Craenmehr, Robert Holwerda et al, "Design Patterns for Mixed-Method Research in HCI," *Proceedings of the 8th Nordic Conference on Human-Computer Interaction: Fun, Fast, Foundational* (October 2014): 361–370.

2. Annie Gentes, *In-Discipline of Design: Bridging the Gap Between Humanities and Engineering* (Springer Nature, 2017).

3. Koen van Turnhout, Marjolein Jacobs, Miriam Losse, Thea van der Geest, René Ronald Bakker, "A Practical Take on Theory in HCI," *White paper* (2019).

What is applied design research?

Applied design research aims to deliver design-relevant knowledge by solving real-world design problems. It is a pluriform research tradition[1] appropriating many approaches from other fields.[2] Still, it is distinctly recognizable by its commitment to improving our designed world. The taxonomy of theory functions by Van Turnhout et al. offers a comprehensive overview of the types of knowledge that are typically connected within a design project.[3]

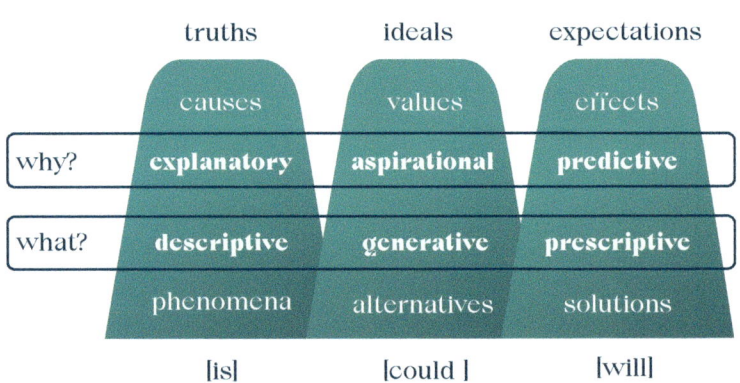

Figure 1
Six well-known theory functions are mapped to the types of knowledge needed to design a solution.

In essence, the taxonomy maps well-known knowledge functions to the activity of designing. To design, we need knowledge about the current situation (the |is| realm), we also need knowledge about desirable alternatives (the |**could**| realm), and about what we know to be effective mechanisms and solutions (the |**will**| realm). Each realm has its own epistemological commitments, and for each type of knowledge one could formulate different criteria for its applicability, as mapped out in Figure 1.

Within this taxonomy, different (design) research traditions could be characterized by highlighting which theory functions they consider to be the most vital outcomes of their discipline. Many social sciences restrict themselves to description and explanation (|**is**|), while some engineering sciences consider prediction and prescription as their key assets |**will**|).

34

In our view, applied design research incorporates all theory functions in the following way.

- |**is**| We commit ourselves to identify and address real-world problems in current society and reframe them, opening up an opportunity space for design. We share this commitment, and many of our approaches, with the Scandinavian school of participatory design.[4]
- |**could**| We consider proposing alternatives and formulating underlying values and ideals in a design vision as our primary contribution to these problems. In this commitment, we feel related to the traditions of speculative design,[5] opportunity-oriented design [6][7] and constructive design research.
- |**will**| We commit ourselves to demonstrate and validate the effectiveness of the alternatives that we propose, preferably by implementing them in current practices. In this commitment, we feel related to design science research.[8][9]

Let's turn to our Sudanese radio station again to apply these commitments in our Radio Dabanga project.

|**is**| *The Dabanga radio journalists, who want to cover the news in a country abroad, are for a large part dependent on citizen journalists, people on the ground who send tips. Whatsapp (and to a lesser extent Facebook messages) are Dabanga's "ear to the ground." However, the journalists are in an information overload situation. Some days the newsroom has to process over 3.000 messages a day, especially during the revolution in 2019. With their small team, they have no means to read all of them, let alone assess them and follow up on them. The staff are agonizing over the fact that they might be missing valuable information, information that might be essential to the people of Sudan, and that they would like to broadcast.*

4. Jesper Simonsen and Toni Robertson eds., *Routledge International Handbook of Participatory Design* (Oxfordshire: Routledge, 2012).

5. Anthony Dunne and Fiona Raby, *Speculative Everything; Design, Fiction, and Social Dreaming* (Cambridge, MA: MIT Press: 2013).

6. Caroline Hummels and Joep Frens, "Designing for the Unknown: A Design Process for the Future Generation of Highly Interactive Systems and Products, in *Proceedings of the 10th International Conference on Engineering and Product Design Education* (Barcelona, September 2008): 204–209.

7. Koen van Turnhout, Stijn Hoppenbrouwers, Paul Jacobs, Jasper Jeurens, Wina Smeenk, and René Ronald Bakker, "Requirements From the Void: Experiences With 1: 10: 100," in *Proceedings of the 3rd Workshop on Creativity in Requirements Engineering* (Essen, 2011).

8. Alan Hevner, Salvatore March, Jinsoo Park and Sudha Ram, "Design Science Research In Information Systems," *MIS Quarterly 28, no. 1* (March 2004): 75–105.

9. Joan Ernst van Aken and Daan Andriessen, eds., *Handboek Ontwerpgericht Wetenschappelijk Onderzoek; Wetenschap Met Effect* (The Hague Boom Lemma Uitgevers, 2011).

10. Aletta Smits, Erik Hekman, Koen van Turnhout, "Ear to the Ground: Using Text Mining to Pick Up All Sudanese Voices for Radio Dabanga," *The EuroIA Conference* (Kraków, September 2020).

11. Gesche Joost, Katharina Bredies, Michelle Christensen, Florian Conradi, and Andreas Unteidig eds., *Design as Research: Positions, Arguments, Perspectives* (Basel: Birkhäuser, 2016): 224.

|**could**| *In the project, we explored whether the tips of the Dabanga journalists could be organized with the help of text mining algorithms. Topic modelling, for example, is a statistical technique that can identify messages with similar topics and group them. This can be used to create a dashboard where journalists see the messages organized by topic rather than chronologically, allowing them to pay attention to less frequent, but important, messages that might not stand out enough to be noticed. We explored these solutions of data processing and visualization in the light of journalistic values such as unbiased overview, scrutiny of information and autonomy of the journalist to act on the provided information.*

|**will**| *The effectiveness of the envisioned pattern classification techniques depends crucially on contextual factors, such as language use in the community we are tailoring to and the modality of the messages (voice text/typed text). Topic modelling is a technique that has been developed and is mostly used for English. Arabic, however, has different linguistic characteristics demanding other pre-processing methods – an Arabic stop word list and a procedure that is specialized in Arabic conjugation patterns – that we needed to apply to the project. Also, citizen journalists inserted code words in messages to throw off potential government spies. These words were unknown to us and complicated the clustering. In other words: applying the clustering in practice yielded knowledge about the contingencies and requirements for using these techniques in practice.*

Six horizons of applied design research.

Who benefits from the knowledge that we develop? Tackling an urgent contemporary and rich design problem like the Dabanga case [10] in all its real-world complexity, sparks several intertwined knowledge agendas. In many ways, each project gives us the feeling that we are only just beginning. No wonder that some authors characterize design research as an indiscipline, a field without standardized methods of developing knowledge, and even propose to commit to this freedom of approach.[11] We share this sentiment to some extent, but unravelling the different horizons of the project will illuminate how we can be anchored in current practices and real-world complexities and simultaneously be pro-grammatic about it, i.e. deliberately working towards knowledge transfer.

Horizon 1: Solving the practical problem (is, could, will).

The straightforward answer to the question 'who benefits' is Radio Dabanga and the people of Sudan. By enabling the journalist to sort messages in a more balanced way and building support for source verification, we empower the Dabanga journalists and increase the quality of news reporting in Sudan. These solutions incorporate integration of |is|, |could| and |will| knowledge restricted in scope to the particular project; it is not easily transferable as an integrated whole.

Horizon 2: Stakeholder's knowledge agendas

The Dabanga project was implemented, among others, in collaboration with Free Press Unlimited, an organization supporting journalists abroad, a journalism research group and a technical partner who built their prototype. Each participated in the project with their own knowledge goals (sometimes overlapping with the knowledge horizons that follow). We feel that successful applied design research projects need to be tailored to the knowledge needs of such partners.

One could argue that the knowledge developed in a project is contingent on the specific project situation; surrendering to such contingencies is a disadvantage of applied design research. This is the case for the integrated solution, but disentangling the different types of knowledge shows that each type of knowledge developed in the project has its own opportunities for transferability.

Horizon 3: Identifying problem families and opportunity spaces (is)

When we look at the |is| knowledge in the project, we see that the problem framing that we developed in the Dabanga project opens up an opportunity space for related problems. Interfaces like the ones developed in the project can help other radio stations broadcasting for their audiences abroad, other journalists in general and other sectors struggling with

37

information overload. Simply put: solving a problem increases its importance [12] for others, and problem framing is a reusable outcome of design research.[13] Similarly, the difficulties of dealing with local language will play out differently in different projects. Nevertheless, we have identified it as an essential attention point for people trying to tackle similar problems.

Horizon 4: Expanding solution repertoire (could)

We now shift our attention from people who may have an information overload problem to the UX designers increasingly working in data-driven ways. We need to sketch how a project like Dabanga benefits them. The simple answer is that solutions are contagious. Van Turnhout and Smits argue that design disciplines are defined to a large extent by the repertoire of solutions professional designers have mastered.[14] Data-driven designers have knowledge of techniques to optimize the seamlessness, engagement and personal relevance of digital media. Dabanga adds scrutiny to this list. Designers faced with a novel design situation can use the Dabanga case as a primary generator of new ideas for solutions in their context even if this is unrelated to the original case.[15] To facilitate this transfer to the professional practice, it is important to explicate the connections from the design to the underlying values (see Figure 1); alternative solutions need to be held accountable to the ideals and values they represent.

Horizon 5: Investigating feasibility: developing the craft (will)

Designers and data scientists can also benefit from the |will| knowledge that we developed: the evidence collected that proves our solution is effective. The many practical problems we needed to solve to get the project up and running can contribute to the development of the craft of data-driven design. Others can reuse the practical tools that we appropriated to deal with local language and can get a feel for the difficulties that may arise in such a project, building to a general feel of feasibility that they need in assessing novel problems.

Horizon 6: Programmatic design and theory annotation.

Most authors on design research argue that it should not consist of independent projects but take place in design programs that have a theoretical core.[16] These programs ensure knowledge build-up across projects. In practice, there is a tension between our commitment to solve real-world problems and programmatic design: arguably, our design projects portfolio is less coherent than that of research groups with a primary commitment to a theoretical core.

However, we do work programmatically, but we focus on theory annotation rather than theory building. Gaver & Bowes[17] correctly observe that design is underdetermined by theory and that theorizing is underdetermined by the designs made in projects like ours. A project like the Dabanga project demonstrates the appropriate and actual use of theories that we use for guidance. As such, the relation between the knowledge we utilize in the project is not hierarchical – the project as an illustration of theoretical insights – but horizontal – the project can be related to theory with a certain amount of subtitling. A portfolio of solutions, connected through such annotations, in turn, also forms a body of knowledge that benefits design.

12. Larry Laudan, *Progress and Its Problems: Towards a Theory of Scientific Growth* (Berkeley, CA: University of California Press, 1978).

13. Donald Schön, *The Reflective Practitioner: How Professionals Think in Action* (New York, Basic Books, 1984).

14. Koen van Turnhout and Aletta Smits, "On Solution Repertoire," in: *Proceedings of the 23rd Engineering and Product Design Education Conference (*Herning, Denmark, 2021).

15. Jane Darke, "The Primary Generator and the Design Process," *Design Studies* 1, no. 1 (1979): 36–44.

16. Ilpo Koskinen, John Zimmerman, Thomas Binder, Johan Redström and Stephan Wensveen, *Design Research Through Practice: From the Lab, Field, and Showroom* (Amsterdam: Elsevier, 2011).

17. Bill Gaver and John Bowers, "Annotated Portfolios," *Interactions* 19, no. 4 (2012): 40–49.

Conclusions and outlook

It turns out that yes, it is possible to design powerful tools that aid journalists in filtering lots of information.

This is good news for other professionals facing similar problems, and for designers and data scientists who'd like to add such solutions to their repertoire. In this paper, we have examined the Dabanga project as an interplay of different types of knowledge: |**is**|, |**could**| and |**will**| knowledge. We also argued that each of these types of knowledge has its own transfer horizon, benefiting different professional groups in different ways. It is this culmination of professional and scientific interests that makes many applied design research projects so rich and unique; but which also poses challenges for programmatic research planning. Understanding how applied research annotates theories and contributes to the solution repertoire of professional designers can aid this planning. We hope to expand on these notions in the coming years, and to illustrate them with more unique, rich projects.

Koen van Turnhout & Aletta Smits

Utrecht University of Applied Sciences

Dr. Koen van Turnhout is professor of Human Experience & Media Design at the Utrecht University of Applied Sciences. The research group focuses on User Experience (UX) professionals. Koen did his doctorate research at the Eindhoven University of Technology with interdisciplinary design research into speech interaction in a social context. His current research is aimed at the methodology of design (research) and the designing of data-driven smart products and services. Koen is the chairman of the Design Science Research Group, a community of practice for design-focused research, and of CHI Nederland, the professional association for human-computer interaction professionals. Dr. Aletta Smits is an associate professor for the Human Experience & Media Design research group. Aletta obtained her doctorate at the University of Amsterdam studying computational linguistics; she is currently researching data-driven user research and user experience design. She has developed the Data-Driven Design master course and, apart from her work at the Utrecht University of Applied Sciences, she is also a public speaker on subjects such as 'how people make choices' and the development of the adolescent brain.

41

Learning from prototypes

From the design studio to the city

Tomasz Jaskiewicz

My recently started Civic Prototyping research group aims to develop new tools and methods enabling urban residents to exploratively research and develop applications of new technologies. This allows them to improve their everyday lives by taking their own initiative in creating valuable services, products, collaborations, and shared spaces. Facilitating applied design research is an essential part of this. But what does applied design research look like in the context of a community of people who keep trying to change the world around them? And what are the challenges for the implementation of applied design research in such a context? To answer these questions, I first need to explain my understanding of what applied design research actually is.

The meaning of applied design research

'Design research both inspires imagination and informs intuition through a variety of methods with related intents: to expose patterns underlying the rich reality of people's behaviors and experiences, to explore reactions to probes and prototypes, and to shed light on the unknown through iterative hypothesis and experiment'. [1] This elegant quote by Jane Fulton Suri perfectly captures my understanding of what is applied design research.

1. Jane Fulton Suri, "Informing Our Intuition: Design Research for Radical Innovation," *Rotman Magazine* (Winter 2008): 52–57.

43

2. For a comprehensive definition and overview, please refer to: Pieter Jan Stappers and Elisa Giaccardi, "Research Through Design," in *The Encyclopedia of Human-Computer Interaction, 2nd edition, eds.* Mads Soegaard and Rikke Friis-Dam (Aarhus, Denmark: 2017): 1–94.

3. William W. Gaver, "What Should We Expect From Research Through Design?," in *Proceedings of the 2012 ACM Annual Conference on Human Factors in Computing Systems (May 2012): 937–946,* https://doi.org/10.1145/2207676.2208538.

4. Abigail C. Durrant, John Vines, Jayne Wallace, Joyce S.R. Yee, "Research Through Design: Twenty-First Century Makers and Materialities," in *Design Issues* 33, no. 3 (Summer 2017): 3–10, https://doi.org/10.1162/DESI_a_00447.

The term 'design research' in academic circles has grown to mean the study of designers, design processes, and their outcomes. Adding the prefix "applied" brings the term back to how it functions in designers' common speak. There, it simply means all kinds of activities that designers do to understand better the context they design for. To me, applied design research means exactly that: the hands-on, practical, but also often informal investigation into the design context, which is an integral part of doing design.

Internationally, the discourse on applied design research and the synonymous term *research through design* [2] has grown considerably over the last five years. Bill Gaver, among others, published an insightful set of challenges for the academic research through design community,[3] and the first Research Through Design (RTD) conference followed in 2015.[4] What made this conference exceptional was its relevance to both academics and design professionals. During the RTD conferences, several styles of applied design research were brought together, and prototypes were used as a valid form of knowledge transfer.

In 2019 we had the honor to host the RTD conference at TU Delft. We saw first-hand how the discourse on applied design research has matured in recent years. However, the diversity of research through design approaches has also given rise to a discussion about what defines 'good' research through design practice, guaranteeing the validity and generalizability of design knowledge.

Zigzagging between design research and design activities

Applied design research can be challenging for designers – simply because research and design are two activities with very different purposes. Research is focused on generating knowledge about the world in which we live. Design is aimed at producing interventions that will change this world. This friction plays out all too often in design processes. Design research focuses on learning about the design context and generating new knowledge. The focus of design is on

applying knowledge to create an intervention that (in the eyes of the designer) will improve the world.

Designing and design researching can be seen as two parallel processes that stimulate each other but have different goals. A good designer iteratively moves back and forth between those two processes, as visualized in Figure 1. In the poetic words of Donald Schön: *"the designer (...) shapes the situation in accordance with his initial appreciation of it, the situation 'talks back' and he responds to the situation's back talk."*

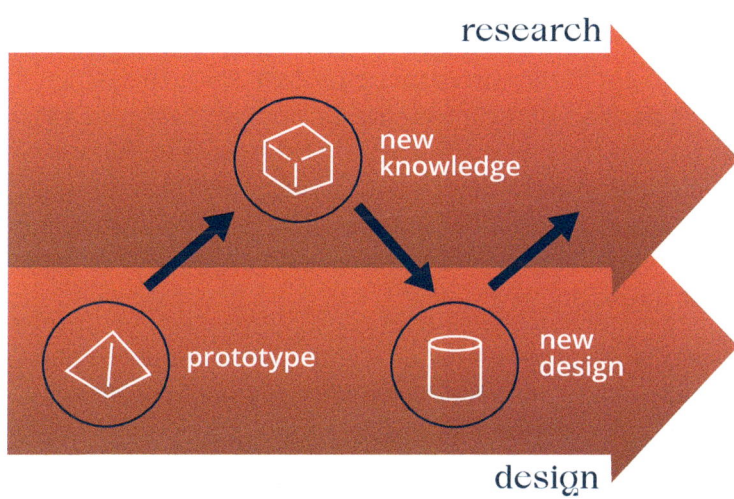

Figure 1
Designers make iterative moves between designing and researching their design context.

Managing one's own design iterations is a difficult skill. I have coached numerous design students who were hopelessly stuck in their design research. They would not dare to come up with any design ideas until their research felt truly complete. This is what design coaches often call *'analysis-paralysis'*. Paradoxically, the more the students researched, the less complete their research felt. At the same time, other students had design ideas in the first moments of their design process and rejected the need for doing design research altogether. They were fixated on their first ideas, and immediately wanted to invest a lot of time and energy in their detailed development. Driven by the loss aversion, they would then do everything they could to protect their 'design darling' from any research or criticism that might prove it flawed.

5. Emil Flach's team-mates were Marieke Noordermeer, Yu Wang and Ward Groutars in the first stage of the project, and Sarah Kraan, Maira Ribelles and Ziwei Li in the second, coached by Roy Bendor and Marise Schot. This story uses the perspective of one person to emphasize the individual character of learning during design.

6. Roy Bendor, Aadjan van der Helm and Tomasz Jaskiewicz, eds., *A Spectrum of Possibilities: A Catalog of Tools for Urban Citizenship in the Not-So-Far Future* (Delft University of Technology, 2018).

My best design students were able to continuously move between their design research and design activities. They kept adapting their ideas, building many prototypes, and gathering feedback on these prototypes from others, making their design research and design development progress work hand-in-hand and support each other.

An example

Let me give you an example to better explain the complexity of applied design research in practice. Emil Flach was a fourth-year Industrial Design Engineering student at the TU Delft when he was commissioned to design a speculative 'instrument of citizenship for Rotterdam 2060'. The assignment was part of the Interactive Technology Design course. At the beginning of the project, Emil and his team [5] were told that they were expected to come up with an application of interactive technology that would help future city dwellers to be more informed, more active and influential in shaping their future city.

Over the next nine weeks, Emil and his team would come up with several ideas about what the future of Rotterdam could hold and invent interactive products that would fit into that future. One of these products was an interactive device that looked like an umbrella and could help create personal space in a busy city (Figure 2). Each of such prototypes helped Emil and his team imagine the future city in more detail and grasp the complexity of future urban problems.

They did so by investigating the city and citizens of today and extrapolating the observed trends into the future. These investigations led them to a future vision of an overcrowded and competitive society. The team converged on a design for a device that would help people to prove themselves as valuable to their community, while at the same time raising many ethical questions about the balance between a person's obligations as a citizen versus personal freedom.

Next, the teams were rearranged. Emil joined his new teammates, and this new cooperation brought up another aspect of overcrowded society, namely dealing with immigrants who settle in the city. The process ultimately led to a

provocative concept of a device that links migrant families to current city dwellers as hosts (Figure 3).

The users of the device still had several choices, but in the end, citizens were always forced to take care of newcomers to their society. The prototype led to much discussion about the validity of the different attitudes that citizens may have toward migrants. It confronted the seemingly noble idea of 'adopting' a migrant family with a forced, automated, and dehumanized way to implement it. It challenged people to question their own values and beliefs about migration.

During the design process, Emil's team's prototypes were shared with other students through work-in-progress exhibitions organized as part of the course.[6] During the project, students regularly exchanged insights and

know-how, helping each other express and understand the complexity of the current and future challenges facing Rotterdam and its inhabitants.

The 'smart migrant dispenser' was later exhibited with other prototypes during the Dutch Design Week in 2018, where it reached thousands of visitors. Emil observed the reactions of people to the prototype while he supervised the exhibition. Many visitors were puzzled. Some laughed, others shrugged their shoulders. Some of them got angry with Emil because they were offended by the taboos that the prototype crossed, or because they misunderstood its thought-provoking purpose.

In his recent reflection on this project, Emil noted that for him personally the key lessons learned during the project were the technical skills he developed while building his prototypes and the ability to explore a design space iteratively. He also regretted that he could not articulate his team's nuanced views on migration when confronted by visitors during the Dutch Design Week exhibition. It was a skill that other team members focused on. Each of the more than a hundred other students who followed the Interactive Technology Design course that year went through a different, very personal learning process. However, they all had the same challenge in mind, improving the future of Rotterdam.

The knowledge of designers

When following the story of Emil's applied design research, you may notice that there is not a single topic or type of knowledge that he has obtained when working on his project. Many of his learnings were tacit and manifested themselves rather in his design actions than in what he said or wrote. Many of his observations, impressions and thoughts probably got lost in between his iterations, while in various ways they still influence Emil's abilities as a designer and design researcher.

In recent years, I have been researching ways to structure, capture, and share practical design knowledge such as Emil's. In a series of studies, my colleagues and I have analyzed the design research documentation of large groups of students. Our analysis revealed three ways to differentiate the different types of design knowledge the students documented, as illustrated in Figure 4.

First, that knowledge was either related to the domain of the design context or the design process. Second, we encountered three different types of knowledge descriptions. There were declarative statements, commonly called 'insights', that described what designers considered to be true. There were also procedural descriptions, which we called 'know-how'. They represented a process needed to be followed to achieve a specific result. There were also speculative state-

ments, which we called 'assumptions', 'design hypotheses' or simply 'ideas'. They described what designers expected a specific intervention to achieve. Third, the topic of the acquired knowledge differed. Some students focused on individual people, others on society at large, or on technology, while in many cases, a combination of topics was addressed.

Figure 4
The three sides of design knowledge detail what designers learn.

This systematization of design knowledge has further helped us to better support applied design research by creating a 'reflection card' tool for structured reflection during the design process. Such a reflection card is a simple digital form organized based on the identified design knowledge categories. Each student designer had to fill in a reflection

7. The statement that 'everyone is a designer' was popularized by IDEO's Tim Brown together with the concept of "design thinking," [8] but it has long existed in the design discourse. For example, Herbert Simon wrote in *The Sciences of the Artificial*: [9] *Everyone designs who devises courses of action aimed at changing existing situations into preferred ones.* And Victor Papanek wrote in *Design for the Real World:* [10] *All men are designers. All that we do, almost all the time, is design, for design is basic to all human activity.* This does not negate the importance of design expertise. In *Design, When Everybody Designs: An Introduction to Design for Social Innovation,* [11] Ezio Manzini clarifies the difference between "diffuse design" performed by non-experts with their intuitive design capacity, and "expert design" which requires trained professionals.

8. Tim Brown, *Change by Design. How Design Thinking Transforms Organizations and Inspires Innovation* (New York: Harper Collins-Publishers, 2009).

9. Herbert Simon, *The Sciences of the Artificial,* Third Edition (Cambridge, MA: MIT Press, 1996).

10. Victor Papanek, *Design for the Real World. Human Ecology and Social Change* (St Albans: Paladin, 1974).

11. Ezio Manzini, *Design, When Everybody Designs: An Introduction to Design for Social Innovation* (Cambridge, MA: MIT Press, 2015).

card for each design or prototype iteration created. In some cases, this meant 20 cards per student. The cards forced student designers to briefly reflect on their design process and its outcomes and articulate the most recently acquired knowledge.

Based on our analysis, we determined that the use of reflection cards involved a sequence of six different activities:

1. Changing the mindset from design-oriented to research-oriented
2. Articulating knowledge
3. Generalizing knowledge
4. Sharing knowledge with others
5. Validating knowledge
6. Applying knowledge to the design

In each of these activities, designers encountered different kinds of challenges. They often were tempted to describe what they did rather than what they had learned. Recording very project-specific notes was also much easier than making more generally applicable statements. However, the effort to articulate and generalize their insights, know-how and ideas proved to be a valuable means of communicating with others. The student designers who were better able to articulate their knowledge gathered more valuable feedback from peers and coaches and could better communicate their project to the outside world.

The articulation of knowledge during the design process also enabled serendipitous connections among students from different teams, sparking collaboration opportunities. Rather than discussing the designs, the student designers began to exchange insights, know-how and ideas more often, turning the design studio into a design research community.

Towards 'civic prototyping'

The challenges faced by the students in our design studio are also at play in cities. In many ways, a grassroots civic initiative, a civic hackathon, or a maker community in many ways resemble an exploratory design studio. The rapid construction of prototypes, articulating and sharing of the accumulated knowledge, communicating and working in multidisciplinary groups are all challenges that such communities face. People in those communities who innovatively try to improve their city are in fact also designers,[7] with unique expertise, insights, and skills. The question remains though, can they all also be design researchers? The articulating and sharing of practical knowledge remains a challenge for both professional and non-professional design researchers. Structured reflection can help, and we can certainly continue developing our tools, methods, and techniques to better support different creative communities.

Tomasz Jaskiewicz

Rotterdam University of Applied Sciences

Dr. Tomasz Jaskiewicz was appointed as a professor at the Creating010 research centre in March 2021, where he leads the Civic Prototyping research theme. Within this theme, he researches new applications and methods, tools, and processes to involve city dwellers in the digital innovation of their social and physical environment. Tomasz has a background in architecture and urban planning and has practical work experience in developing experimental architectural projects, interactive installations, and digital design tools. In 2013, he obtained his PhD from the Faculty of Architecture at the Delft University of Technology. From 2014, he worked as an assistant professor at the Faculty of Industrial Design, where he currently holds a design fellow position.

Design thinking for professionals

Applied design research as a driving force for innovating education

Guido Stompff

In their distinct practices, professionals sooner or later face problematic situations where they do not know what to do. They are faced with issues they haven't experienced before and for which no ready-to-use solutions are available. Therefor they can not rely on experience. For instance, a communication professional discovers that waste separation information campaigns are failing because plastic waste contains much more residual waste than expected. Or an IT administrator is confronted with a new type of virus and needs to respond quickly. In such situations, professionals should not only be able to explore what the problem is, but also to come up with new solutions.

To achieve this, professionals need next to research abilities and also some design capacities. Designing is devising plans of action to turn an existing, problematic situation into a more preferred situation.[1] So, it is not just designers and architects who are designing. Every professional, such as a physiotherapist or a facility manager, occasionally designs. Unfortunately, most professionals are hardly trained in

1. Herbert Simon, *The Sciences of the Artificial, Third Edition* (Cambridge, MA: MIT Press, 1996).

53

2. Horst W.J. Rittel and Melvin M. Webber, "Dilemmas in a General Theory of Planning," *Policy Sciences*, 4, no. 2 (1973): 155–169.

3. Roger L. Martin, *The Design of Business: Why Design Thinking Is the Next Competitive Advantage* (Boston, MA: Harvard Business Review Press, 2009).

4. Richard J. Boland, and Fred Collopy eds., *Managing as Designing* (Stanford, CA: Stanford Business Books, 2004).

5. Tim Brown and Jocelyn Wyatt, "Design Thinking for Social Innovation," *Development Outreach* 12, no. 1 (2010): 29–43, https://doi.org/10.1596/1020-797X_12_1_29.

6. Guido Stompff, *De Kracht van Verbeelden, Design Thinking in Teams,* Inaugural speech (Amsterdam: Hogeschool Inholland, 2020).

7. Kees Dorst, "The Core of 'Design Thinking' and Its Application," *Design Studies* 32, no. 6 (2011): 521–532, https://doi.org/10.1016/j.destud.2011.07.006.

8. Joan Ernst van Aken and Daan Andriessen, eds., *Handboek Ontwerpgericht Wetenschappelijk Onderzoek; Wetenschap met Effect,* (The Hague: Boom Lemma Publishers, 2011).

design. If they have to design, they will opt for an analytical approach, whereby research should lead to insights into the problem and, hopefully, to ideas for a design, or at least design requirements. In practice, this proves to be ineffective: stakeholders respond not as expected, well-intentioned solutions lead to additional problems, it is unclear what is 'good' enough to stop, and unexpected developments change the problem completely. In other words, the problem is *wicked.*[2]

Design thinking seems to fit better with such situations. It was somewhat tautologically defined as 'solving problems the way designers solve problems',[3] but with a substantially different context, such as management [4] or social issues.[5] Design thinking requires empathy, the ability to put yourself in the shoes of the other; it requires creativity to overcome existing dilemmas and expressive skills to portray your ideas.[6] Design thinking is integrating design and research and aptly can be named design research or design-based research. It advances through learning by creating and reflecting on the outcomes, using a different logic (abduction [7]) than the classical sciences.

Design thinking in higher education

Despite the growing interest in design thinking, the development of design ability is hardly embedded in Dutch higher education, although 'ontwerpgericht onderzoek' doing research *a priori* design activities has for some time been heralded as an approach that may establish a scientific way of designing.[8] But, as Daan Andriessen mentions elsewhere in this publication, 'designing itself is hardly used as a means of gaining insight'. Design thinking and design research put the designing at the center. By devising new concepts, creating prototypes, and reflecting on the results, new insights are acquired.

It is time to better position design thinking in higher education. We are facing two problems here. Firstly, the transformation of design thinking to other contexts is still in full swing. Although some design methods, such as customer journey maps, are already widely applied in other contexts, the question remains whether *unaltered* design thinking – as designers do it – is effective for those other contexts. For example, should managers be able to draw? Are energetic hackathons suited to solve social problems? What is a suitable prototype to test a new way of organizing?

Secondly, there is a gap between three different bodies of knowledge: (1) design, (2) all other professions (ranging from physiotherapy to facility management) and (3) the education of professionals. These three bodies of knowledge overlap (see Figure 1): the overlap between design and other professions concerns the designing skills that every professional needs (design thinking). The overlap between education and design is related to how designers are trained, with much attention to creating, creativity and expression. The third overlap is of less importance here. The research gap is about developing design skills for (future) professionals *who are not designers*.

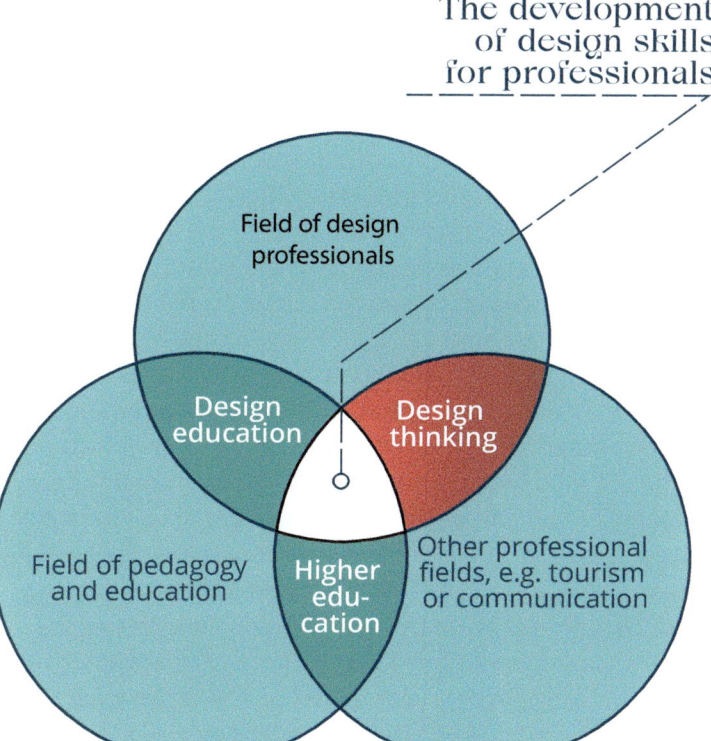

The research gap
The development
of design skills
for professionals

Field of design professionals

Design education

Design thinking

Field of pedagogy and education

Higher education

Other professional fields, e.g. tourism or communication

Figure 1
The research gap of the study on teaching design thinking to non-designers.

55

9. Richard Buchanan, "Wicked Problems in Design Thinking," *Design Issues* 8, no. 2 (Spring, 1992): 5–21.

10. Jeanne Liedtka, "In Defense of Strategy as Design," *California Management Review* 42, no. 3 (2000): 8–30, https://doi.org/10.2307/41166040.

11. Simon, *The Sciences of the Artificial.*

12. See also the contribution by Koen van Turnhout and Aletta Smits in this publication.

Those who have not learned to express their ideas in inspiring ways to stir others, who have not learned how to make and test a prototype, who have not learned that there are dozens of ways to fuel creativity, and who are not used to tackling a wicked problem.

Design research: The development of design skills for professionals

The Creative Business research group at Inholland is linked to various courses, such as Tourism Management, Business Innovation, and Creative Business. These are not classical design courses, but the vision is to also teach the students of these courses relevant design skills. But how do you develop design capabilities in prospective professionals who do not take design courses? Initially, this seemed to be mainly an activity to support education, but gradually it became clear that this is a complex question that requires thorough research. The coaches involved raised all kinds of sub-questions: Which design skills are desired? How do you develop them effectively? How do you assess these skills? Which methods do we teach students? How do we adjust the curriculum? How do we professionalize lecturers?

Noblesse oblige: the study *Educating professionals in design thinking* is set up as a design research project. Design research combines research and design: the interpretation of the problem and creation of possible solutions. Unfortunately, design research is also a problematic union between two fundamentally different worlds, a subject that design theorists have been discussing for decades.[9][10][11] Research focuses on interpreting the existing situation (*what is?*). In contrast, design is focused on what does not yet exist, the future. And within that, design is imaging what is desired (*what might be?*) and designing plans to get there (*what can be?*).[12] This difficult combination leads to many design researchers choosing one or the other. Either they choose *research for design,* with the emphasis on interpreting the existing situation as input for a design. Or they select *research through design*, with the emphasis on imaging the desired and creating and testing designs as input for research.

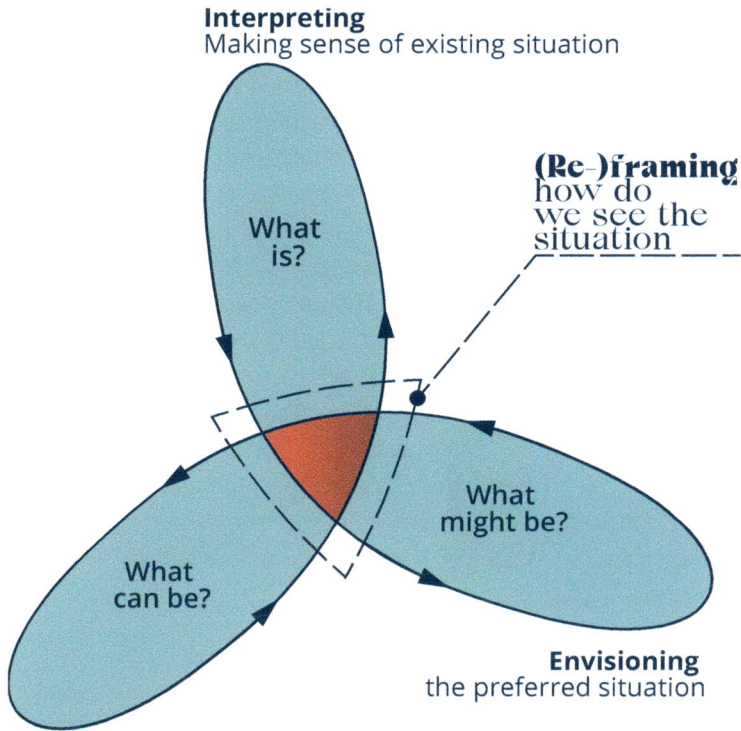

Interpreting
Making sense of existing situation

(Re-)framing
how do
we see the
situation

What is?

What might be?

What can be?

Envisioning
the preferred situation

Design
devising plans of action to change
an existing situation into a more preferred situation

Figure 2
Design research combines different types of activities, which do not take place so much in succession, but iteratively, with the results of the activities influencing the other activities. Some activities are aimed at interpreting the existing situation (sense-making), some at imaging the desired situation (envisioning) and some at designing plans to get from the existing to the desired situation, given the constraints (design). All activities are framed by a frame that develops gradually: how do we see the situation?

However, this practical research requires both research *for* design and research *through* design. Research *for* design is needed to understand what is going on, what the problem is. Lecturer-researchers, therefore, carry out studies on the success and failure factors of adopting design thinking in education. For example, they discovered that coaches with a classical academic background overestimate the value of theoretical frameworks and underestimate the importance of testing. They also learned that the design vocabulary ('concept', 'prototype') is ambiguous and creates much confusion. Research *through* design is needed to create designs and test prototypes. A team of lecturer-designers has been brought together to develop online and offline resources, methods, and learning experiences that can be used for the relevant training programs.

13. Tonnie van der Zouwen, *Actieonderzoek Doen: Een Routewijzer voor Studenten en Professionals* (Amsterdam: Boom Publishers, 2018).

14. Dalila Cisco Collatto, Aline Dresch, Daniel Pacheco Lacerda, Ione Ghislene Bentz, "Is Action Design Research Indeed Necessary? Analysis and Synergies Between Action Research and Design Science Research," *Systemic Practice and Action Research*, 31, no. 3 (2018): 239–267, https://doi.org/10.1007/s11213-017-9424-9.

Applied design research: In and with educational practice

One interesting aspect of this practical research is that it relates to the practice that the researchers (and myself) are a part of! In this sense, it is very similar to action research, which initiates change by researching *in* and *with* the practical environment.[13] In action research, after extensive research and with knowledge of similar situations, interventions are devised and carried out. The outcomes of those interventions are then reflected on. Our research approach combines design research (in the broad sense) with action research. It is research for design and research through design, and is carried out in practice and with practitioners. This generates new knowledge, both in the form of explicit knowledge and in the form of knowledge embedded in methods, artifacts, and tools.

Such research is sometimes called action design research, but the explicit mixing of two research traditions raises the necessary questions.[14] I call this applied design research, because it is an applied form of design research, in which the practical value comes first. It is first and foremost about better education; scientific publications come second. And that choice has consequences because practice is ambiguous, complex, and full of surprises.

Thus, this project regularly cuts corners to respond to unexpected problems. Plans are derailed by additional conditions or when new opportunities present themselves. Practice requires a research design in which agility is crucial to achieve the desired practical goals. For example, it became clear that educational operations raise many practical barriers for experiments. Time schedules that have been in place for months and strict education and examination regulations limit the necessary experimental space. Improvisation skills are a necessity to test new concepts in existing programs. Partly for this reason, there is an intensive collaboration with experienced lecturers in a learning network.

The challenges for applied design research

Applied design research is efficient and innovative because something new is created that transcends existing, familiar solutions. Something new implies new knowledge, possibly relevant for the scientific body of knowledge. Unfortunately, its exploratory nature is at odds with existing academic standards, such as methodological rigor, extensive literature review, data transparency and repeatability of results. Does this mean that applied design research cannot be scientifically justified? No, but it is not easy. I think there are two significant challenges.

First of all, scientific standards are embedded in specific philosophical traditions, including theories on what 'knowledge' is and what 'truth' is. Practice-led research – and not just applied design research! – has a complicated relationship with the specific traditions most sciences are based on, because much of the knowledge is embedded *in practice* (i.e. in the practical environment). It is incorporated in what

Figure 3
A workshop with lecturers and staff of Inholland in the context of educational innovation.

59

professionals do, in the artifacts they create, in the roles they assume, in the tools they use, in the environment where they work. Therefore, scientists who carry out practice-led research place great emphasis on the *tacit dimension of knowledge*,[15] the non-verbalizable form of knowledge. Applied design research can better follow the example of the practice-led sciences, where pragmatism generally permeates. In these traditions, thinking cannot be seen separate from doing, theory not separate from practice, and behavior not separate from the environment. It provides a rich basis for understanding design and applied design research, but although there is a renaissance of pragmatism,[16][17] the translation to scientific standards for applied design research has not yet been done sufficiently.

Figure 4
Co-creation workshop
with lecturers and staff of
Inholland.

A second challenge is the experience of researchers applying applied design research. Experience plays a significant role in design: senior designers can successfully complete complex assignments that contain so much uncertainty and ambiguity that inexperienced designers do not know how or where to start. This means that researchers without design experience are poorly equipped to carry out projects in which design activities play an essential role. Applied design research requires design expertise. The challenge is how we teach researchers these skills. Or, conversely, how we teach experienced designers research skills. In short: applied design research places high demands on those who apply it!

15. Michael Polanyi, *The Tacit Dimension* (New York: Doubleday Anchor, 1966).

16. Peter Dalsgaard, "Pragmatism and Design Thinking," *International Journal of Design* 8, no. 1 (2014).

17. Brian Dixon, *Dewey and Design: A Pragmatist Perspective for Design Research* (London: Springer Nature, 2020).

Guido Stompff

InHolland University of Applied Sciences

Dr.ir. Guido Stompff has been a Professor of Design Thinking at the Creative Business research group of the Inholland University of Applied Sciences since 2019. After his training as an industrial designer (TU Delft), he worked for over 25 years as a designer, covering the full scope of the field, including product design, UX design, communication design, packaging design, branding, and even art. Since 2003, he has combined his work with teaching at various universities and universities of applied sciences. In 2011, he obtained his PhD in the facilitation of innovation in multidisciplinary teams, leading to various publications on team design and the importance of imagination for innovation processes. His book *Design thinking, radicaal veranderen in kleine stappen* was published in 2018. The book was voted Dutch management book of the year.

Dance? Dance!

The contribution of practice-driven design research to the ballet of disciplines

Peter Troxler

This article invites applied design research to a new challenge: the "dance of the disciplines." It begins with an attempt to *discern* the elements of applied design research. It continues with a set of examples to *display* the relevance of these elements. And it concludes with a call to reclaim *design* and reintroduce it in transdisciplinary practice through boundary crossing, the "dance of the disciplines."

Discern

Design, since Archer [1] and Cross,[2] is understood respectively as its own "discipline" and an "area of education" – distinct from the two more traditional areas of education in science on the one hand, and arts and humanities on the other – with its own study subject, its own goals, values, and methods:

- Design studies the man-made world, whereas science studies the natural world and the humanities the human experience.
- Design seeks appropriateness, while science seeks truth and humanities justice.

1. Bruce Archer, "Design as a Discipline," *Design Studies* 1, no. 1 (1 July 1979): 17–20. https://doi.org/10.1016/0142-694X(79)90023–1.

2. Nigel Cross, "Designerly Ways of Knowing," *Design Studies* 3, no. 4 (1982): 221–27. https:// doi.org/10.1016/0142-694X(82)90040–0.

3.   Merriam-Webster. Research. In *www.merriam-webster.com*, retrieved 31 January 2021, from https://www.merriam-webster.com/dictionary/research.

4.   OECD, *Frascati Manual 2015: Guidelines for Collecting and Reporting Data on Research and Experimental Development, the Measurement of Scientific, Technological and Innovation Activities* (Paris: OECD Publishing, 2015).

5.   Daniel Fallman and Erik Stolterman, "Establishing Criteria of Rigor and Relevance in Interaction Design Research," *Proceedings of Create10 – The Interaction Design Conference* (2010).

6.   Peter Miller, "Reliability," in *The SAGE Encyclopedia of Qualitative Research* (Thousand Oaks, CA: SAGE Publications, Inc., 2008): 753–754.

7.   Peter Miller, "Validity," in *The SAGE Encyclopedia of Qualitative Research* (Thousand Oaks, CA: SAGE Publications, Inc., 2008): 909–910.

8.   Ministerie van Binnenlandse Zaken en Koninkrijksrelaties, Wet op het Hoger Onderwijs en Wetenschappelijk Onderzoek, retrieved 9 February 2021, https://wetten.overheid.nl/BWBR0005682/2021–01–01.

9.   Vereniging van Hogescholen, *Brancheprotocol Kwaliteitszorg Onderzoek* (October 2015).

- Design values practicality, ingenuity, and empathy. In contrast, the values in science are objectivity, rationality, and neutrality; in the humanities, they are subjectivity, imagination, and commitment.
- Design methods are modelling, pattern-formation, and synthesis; science methods are controlled experiment, classification, and analysis; humanities methods are metaphor, criticism, and evaluation.

Research is understood to be the production of knowledge, the

> *studious inquiry or examination, especially investigation or experimentation aimed at the discovery and interpretation of facts, revision of accepted theories or laws in the light of new facts, or practical application of such new or revised theories or laws.* [3]

Such "studious inquiry" is "creative and systematic."[4] *Creative* implies that research can produce new findings based on original, not obvious concepts or hypotheses, of which the final outcome is uncertain. *Systematic* requires research to be conducted in a planned way that documents the steps taken and the outcomes achieved. Systematic also implies that the results could be possibly reproduced and used elsewhere. Systematic research is rigorous and relevant.[5] Typically, systematicity of research is operationalized as validity and reliability of research. Validity stands for the "goodness" or "soundness" of a study. However, it cannot be expressed in global criteria except in quantitative, positivist research tradition, so it has to be described according to the purpose and methods of a given design study.[6] Reliability is the "dependability," "consistence" and often "repeatability" of a study. Again, the diversity in many areas of research requires a case-by-case approach.[7]

Applied (with regards to research) signifies that applied design research is prompted by and feeds straight back into professional design practice. In the Netherlands, there is a legal difference between "academic research," carried out at universities, and "research directed at professional practice" at universities of applied sciences. This distinction is instrumental for the distribution of government funding for research – namely only for research at universities.[8] As such, applied research is much closer to the actual applications of the new knowledge it produces, thus requiring little extra "technology transfer" efforts. Applied design research is *"rooted in professional design practice. Applied design research is prompted by professional design practice (real-life situations), in both profit and non-profit sectors. Applied design research generates knowledge, insights and products that contribute to solving problems in professional design practice and developing this practice."*[9]

Applied design research, therefore, is the creative and systematic production of knowledge that is prompted by and feeds right back into shaping the man-made world through modelling, pattern-formation, and synthesis, achieving appropriate results that can be tested for practicality, ingenuity, and empathy.

Display

In terms of **applied,** applied design research garners its relevance from focusing on the "applied" in design research – and this is not just semantics. Applied design research takes its cues from design practice, not only from the tribulations of mainstream practice but from the fringes where investigation and experimentation are required to develop the technique.

Figure 1
Open Design: Demonstration of the open-source Wiki House in Vienna, 2015. Photo: © 2015 Claudia Garad (cc-by-sa), retrieved from https://commons. wikimedia.org/w/index. php?title=File:Wiki-House_Wien_Eröffnung_ II.jpg&oldid=493867214

10. Bas van Abel, Roel Klaassen, Lucas Evers, Peter Troxler, *Open Design Now: Why Design Cannot Remain Exclusive* (Amsterdam: BIS Publishers, 2011).

11. Peter Troxler, "The Beginning of a Beginning of the Beginning of a Trend," in Bas van Abel, Roel Klaassen, Lucas Evers, Peter Troxler, *Open Design Now: Why Design Cannot Remain Exclusive* (Amsterdam: BIS Publishers, 2011).

12. Peter Troxler and Patricia Wolf, "Look Who's Acting! Applying Actor Network Theory for Studying Knowledge Share in a Co-Design Project, *International Journal of Actor-Network Theory and Technological Innovation* 7, no. 3 (2015): 15–33.

13. Patricia Wolf and Peter Troxler, "Community-Based Business Models: Insights From an Emerging Maker Economy," *Interaction Design and Architecture(s)* 30 (2016): 75–94.

14. Peter Troxler, "Building Open Design as a Commons," in Loes Bogers and Letizia Chiappini, *The Critical Makers Reader: (Un)Learning Technology*, (Amsterdam: Institute of Network Cultures, 2019): 2018–226.

15. Roland Jochem, "The Future of Product Creation is Open and Community-Based," *Research Outreach* 113 (2020): 6–9, https://doi.org/10.32907/RO-113-69.

16. Robert Anderson, *European Universities From the Enlightenment to 1914* (Oxford: Oxford University Press, 2004).

Figure 2
Designing Education: Manon Mostert – van der Sar (right) working with educators in Utrecht, 2019. Photo: Roy Borghouts.

For example, when the concept of open design emerged,[10] [11] designers were baffled by the idea that they could seriously be required to even think of relinquishing the business model they believed they were thriving on – earning royalties on their "intellectual property." Through investigation [12] [13] and experimentation [14] – which is still ongoing [15] – researchers tried to approach the phenomenon, understand the frictions and develop ways to "do open design."

In terms of **design,** applied design research garners its relevance from focusing on design approaches in applied research. As such, it can be related to design as the practice of design professions and their development – architecture, landscape, furniture, fashion, light, product, package, graphic, web and so forth, or it can be related to design in other practices, such as organization design, research design, and education design, and studying the contributions of design by organizing, researching or teaching the subject of the man-made world.

For example, many educators express the view that with the teaching methods and school systems – many of which stem from the late 19th and early 20th century [16] – they are insufficiently equipped to teach in the supposedly VUCA [17] environment of the early 21st century.[18] A design approach to educational processes, devised by a designer, developed by teachers, and encouraged by school administrators, promises to "transform education on a small scale but with a big impact."[19]

In terms of **research,** applied design research garners its relevance from creatively and systematically producing new knowledge in the field of design – using designerly methods such as modelling, pattern-formation, and synthesis, also known as "abduction" [20] – and validating the research results by ascertaining that they are useful in a given practice, that they are clever and original for that practice, and that they are sensitive towards and in rapport with the practice.

For example, after maker spaces in libraries started to emerge,[21] the National Library of the Netherlands wanted to investigate if there was indeed a way forward for this new concept as part of their digital strategy. A design research project was set up that studied that question and came up – through several design sessions – with a roadmap, outlining three lines of development – policy development, curriculum development, and community development.[22] This roadmap was then validated.

Design

Design and design research (applied or not) have come a long way since emerging from engineering and "trying to bend the methods of operational research and management techniques to design purposes."[23] The past five decades have seen design growing into a discipline of its own – in education, as a profession – eventually a reflective one.[24] Endless discussion ensued about distinctive definitions of

17. VUCA is an acronym which stands for Volatility, Uncertainty, Complexity, Ambiguity.

18. Ken Robinson, *Out of Our Minds, the Power of Being Creative* (Hoboken NJ: Wiley, 2011).

19. Manon Mostert – Van der Sar, *Hey Teacher, Find Your Inner Designer* (Amsterdam: Boom Publishers, 2019).

20. Lauri Koskela, Sami Paavola, Ehud Kroll, "The Role of Abduction in Production of New Ideas in Design," in Pieter E. Vermaas and Stéphane Vial (Eds.), *Advancements in the Philosophy of Design* (Springer International Publishing, 2018): 153–183, https://doi.org/10.1007/978-3-319-73302-9_8.

21. Theresa Willingham and Jeroen De Boer, *Makerspaces in Libraries, Library Technology Essentials 4* (Lanham MD: Rowman & Littlefield Publishers, 2015).

22. Peter Troxler, Eva Visser and Maarten Hennekes, *Roadmap Makerplaatsen. Van Knutselen 2.0 Naar Leren met 21ste Eeuwse Vaardigheden,* (Rotterdam: Kenniscentrum Creating 010, 2018).

23. Bruce Archer, "Design as a Discipline," *Design Studies* 1, no. 1 (1 July 1979): 17–20. https://doi.org/10.1016/0142-694X(79)90023-1: 17.

24. Donald Schön, *The Reflective Practitioner: How Professionals Think in Action* (New York, Basic Books, 1984).

Figure 3
Validating the roadmap for library maker spaces with (from left to right) library, maker and space professionals at the National Library, The Hague, 2018. Photo: © 2018 Peter Troxler.

25. Nigel Cross, "From a Design Science to a Design Discipline: Understanding Designerly Ways of Knowing and Thinking," In *Design Research Now: Essays and Selected Projects*, ed. Ralf Michel (Basel: Birkhäuser, 2007): 41–54. https://doi.org/10.1007/978-3-7643–8472-2_3.

26. Harold G. Nelson and Erik Stolterman, *The Design Way, Second Edition. Intentional Change in an Unpredictable World* (Cambridge, MA: The MIT Press, 2012).

27. Modified from Nelson and Stolterman, 2012, p. 243, modifications shown as emphasis.

28. Holger Rhinow, Eva Köppen, Christoph Meinel, "Design Prototyps as Boundary Objects in Innovation Processes," in *Proceedings of the Design Research Society International Conference* (Bangkok, July 2012): 1581–1590.

29. Lucy Suchman, "Working Relations of Technology Production and Use," *Computer Supported Cooperative Work* 2, no. 1 (1994): 21–39. https://doi.org/10.1007/BF00749282.

research about, through, from, in, and for design, their subtle nuances enshrined in the programs of conferences, the editorial lines of journals, and professional societies' collective identities.[25] More variations and deep thoughts became the material for myriads of introductory chapters of PhD theses all around the globe.

It is time for design research professionals to leave that discussion there – for that is where it belongs – and move on and continue to actually do design research, in extension to how Nelson and Stolterman [26] summarize what designers do:

"Design researchers *are heavily invested in* understanding, *developing and using good design processes and realizing desired outcomes. In design inquiry, the process is aimed by design intention – desiderata* and new knowledge. *The right process going in the right direction will reach the right outcome*, both in products and the knowledge enshrined in them and distilled from the process. *In other words, desired outcomes are made visible* and communicable *and are successfully achieved with mindful, intentional aiming. Process and outcome are entwined and equally important to the designer* and the design researcher. *A good process, properly aimed in the right direction, reveals the answer to the question: What* (about) *design is desired to be made real?"* [27]

Two things happened in the past years to design that are essential signals that design and, with it, design research, have to move on and move differently. One, design got highjacked by management consultants as "design thinking." Designers need to reclaim design thinking as their professional way to make the desired reality. Two, designers like to understand their discipline as ultimately interdisciplinary – the idea that a designer has insights in all the disciplines (or can gain that quickly) and solves all their problems. This is an attitude that is not so different from that of management consultants, and is often a sign of blissful ignorance or, worse, offensive arrogance. Designers need to return to design to make the desired real.

Making the desired real, however, is not a solo discipline. Designers know how to use artifacts as boundary objects [28] to bring different parties and stakeholders together to develop a common language about the desired outcome, the object of a collective process of problem-solving, of delivering desiderata – the "what" in design. In working together with other disciplines in the entire design process, however, boundary objects alone are insufficient. Differences in how and why disciplinary practices are performed become evident and need to be addressed. Fruitful collaboration in such a transdisciplinary practice emerges in a development called "boundary crossing." [29]

Figure 4
Research as visualised in Violeta Clemente, Katja Tschimmel & Fátima Pombo, "A Future Scenario for a Methodological Approach applied to PhD Design Research. Development of an Analytical Canvas," *The Design Journal* 20 (September 2017): 792–802, https://doi.org/10.1080/14606925.2017.1353025.

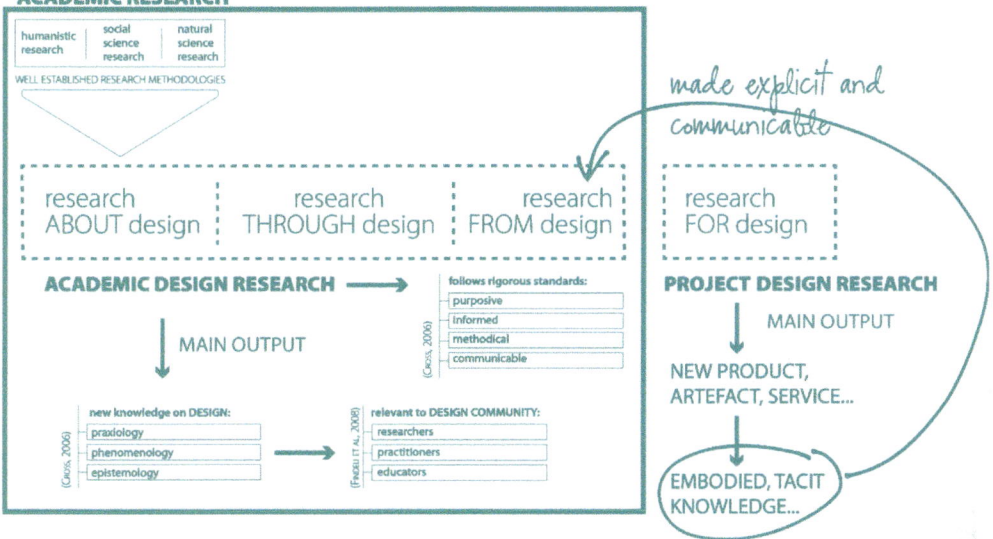

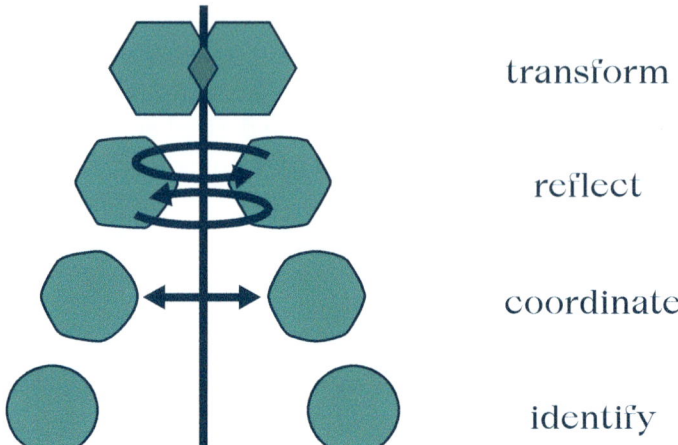

transform

reflect

coordinate

identify

Figure 5
Boundary crossing in
transdisciplinary work.

Boundary crossing is the dance of the disciplines (...) to the rhythm of:

(a) identifying the communalities and differences by making them explicit;

(b) coordinating the collaboration to primarily appreciate and to efficiently address the problem at hand, from what forms a common understanding;

(c) reflecting the differences and reframing the viewpoints based on the multiplicity of perspectives; and

(d) transforming and bridging those different perspectives so that they lead to the new solutions that would not have been possible without the transdisciplinary collaboration.

Boundary crossing is creating power and direction from the communalities and forming new ideas from the generative combination of the difference.[30]

Boundary crossing in transdisciplinary work is the new challenge for applied design research:

- What are designerly ways of knowing in transdisciplinary settings?
- What are design practices and processes of boundary crossing?
- What are the artifacts (form and configuration) that foster boundary crossing beyond the "what"?

It is important to answer these questions from within design and gradually interweave them with the answers from other disciplines to deepen the insights into the contribution of design and communicate it to the outside.

30. Mortaza S. Bargh and Peter Troxler, "Digital Transformations and Their Design – Renewal of the Socio-Technical Approach," in *Hoger Beroepsonderwijs in 2030. Toekomstverkenningen en Scenario's vanuit Hogeschool Rotterdam*, eds. Daan Gijsbertse, Arjen van Klink, Kees Machielse, Jeroen Timmermans (Rotterdam: Hogeschool Rotterdam Uitgeverij, 2020): 326–369.

Peter Troxler
Rotterdam University of Applied Sciences

Dr. Peter Troxler is professor of Revolution in Manufacturing at Rotterdam University of Applied Sciences. He obtained his PhD at the ETH Zurich, at the cutting edge of occupational psychology and business administration, specializing in organisation design. He has worked as a management consultant at a design consultancy firm in Switzerland (1997–2018), as a research manager in artificial intelligence at the University of Aberdeen, Scotland (2001–2004), and as a senior project manager and freelance executive editor at Waag in Amsterdam (2007–2010). He was also the founder, mentor and inspirator for many Fab Labs in Europe (2009–2013). He worked as a producer for an independent theatre group in Switzerland (1994–2001), and was the director of a critical artistic research collective in Aberdeen (2003–2007).

PART 2:
THE URGE
TO IMPROVE
THE WORLD

"If you want truly to understand something, try to change it."

~ Kurt Lewin

Idealistic visions of the future or realistic solutions?

Baby steps towards innovation leaps

Peter Joore

Do I choose idealism or realism? Do I select a training program that allows me to work on undefined dreams, or do I opt for a practical program? Without really knowing what I was getting myself into, in 1985, I went to study Industrial Design Engineering in Delft mainly because, besides the technology, this program also had a creative component. To be honest, back then, a 'higher' design goal to improve the world was unheard of. Designs were aimed at fulfilling a 'function', and ideals were aimed at a smart technical concept, a firm cost price, and a good realization of the ergonomic framework conditions. Of course esthetics also played a role somewhere in the background, although at the end of the day, form always played second fiddle to function.

Form follows function was the adage, and the client's purpose was the deciding factor at all times. That client was almost always an industrial company – the program was named *industrial* design for a reason – looking to sell as many products as possible and making as much profit as possible. The research at the faculty was mainly functional and technically oriented at the time, with the researcher dropping a vacuum cleaner a thousand times to determine after how many times the plastic started breaking. Undoubtedly very important, but not something that inspired me personally.

Halfway through the program, I decided to go on a personal quest for idealism. I took a break and did several things, including working in the shanty towns of Bombay in India for several months. After my time at university, I tried to keep going in that direction, among other things through a not very successful attempt to work as a missionary and development worker in Albania, but eventually, I started working as a product designer. One of my most memorable projects was the development of the new Chek Lap Kok airport in Hong Kong and the stations of the new MTRC subway there, in collaboration with NKI Group, Springtime Design, Total Design, and Norman Foster's architectural firm in London. Not very idealistic, but very challenging and exciting.

Innovation Leaps

The next stage in my development came when I switched to a research group at TNO in Delft. There, the Kathalys research group was working on the development of sustainable system innovations.[1] These innovation leaps focused on a factor 4 sustainability improvement, with the underlying reasoning that if we want to cut the environmental impact of used materials and energy by half, while the population grows and possibly doubles, the ecological impact of a product must therefore be reduced by two-times-two-is-four.[2] This factor 4 was later replaced by a factor 10, which required even more radical innovations.

We soon discovered that if you want to achieve such radical innovation leaps, it is not enough to innovate at the product level alone. To have a real impact, it is necessary to innovate at the level of the product-service system, or rather at the level of the socio-technical system, where different actors each fulfill their role and pursue specific interests. In this kind of innovation, it ultimately proved essential to think carefully about the underlying world vision held by the actors involved. This overarching world view turned out to be a decisive factor for the choices made within innovation projects. This insight provided a direction to connect the idealistic perspective I was still looking for with my work as a designer. This was reinforced by the cooperation with Professor Ezio Manzini of the Politecnico di Milano in the European HiCS research project, which for me was the first time that I met someone who really looked at the design profession from a broader philosophical perspective.[3] [4]

Experimenting with a new mobility concept

One example of such an innovation leap project was a collaborative project where we worked at TNO with partners such as Gazelle, Nike, and Achmea on a mobility concept for individual short-distance transport, called MITKA (Mobiliteitsconcept-voor-Individueel-Transport-op-de-Korte-Afstand, see Figure 1). Our goal was to motivate people to leave the car and to start using a compact electric-driven transport system. At some point, however, it became commercially much smarter to promote the developed vehicle as a trendy off-road cross-vehicle for affluent yuppies. Although that may have been smart from an economic point of view, in terms of our ideals, it was the opposite of what we had intended. Therefore, we soon discarded that option.

1. Adrie Beyen, *Kathalys: Vision on Sustainable Product Innovation* (Amsterdam: BIS Publishers, 2001).

2. Ernst von Weizsäcker, Amory B. Lovins, L. Hunter Lovins, *Factor Four: Doubling Wealth, Halving Resource Use* (London: Earthscan Publications Ltd, 1998).

3. Ezio Manzini, Luisa Collina, Stephen Evans, *Solution Oriented Partnership. How to Design Industrialised Sustainable Solutions* (Cranfield: Cranfield Publishers, 2004).

4. François Jegou, Peter Joore, *Food Delivery Solutions* (Cranfield: Cranfield Publishers, 2004).

5. Peter Joore, Michel van Schie, Eindrapportage MOVE – Mobiliteitsconcept voor Individueel Transport voor de Korte Afstand – MITKA (Delft: TNO, 2001).

Figure 1
MITKA: mobility concept for individual short distance transport.[5]

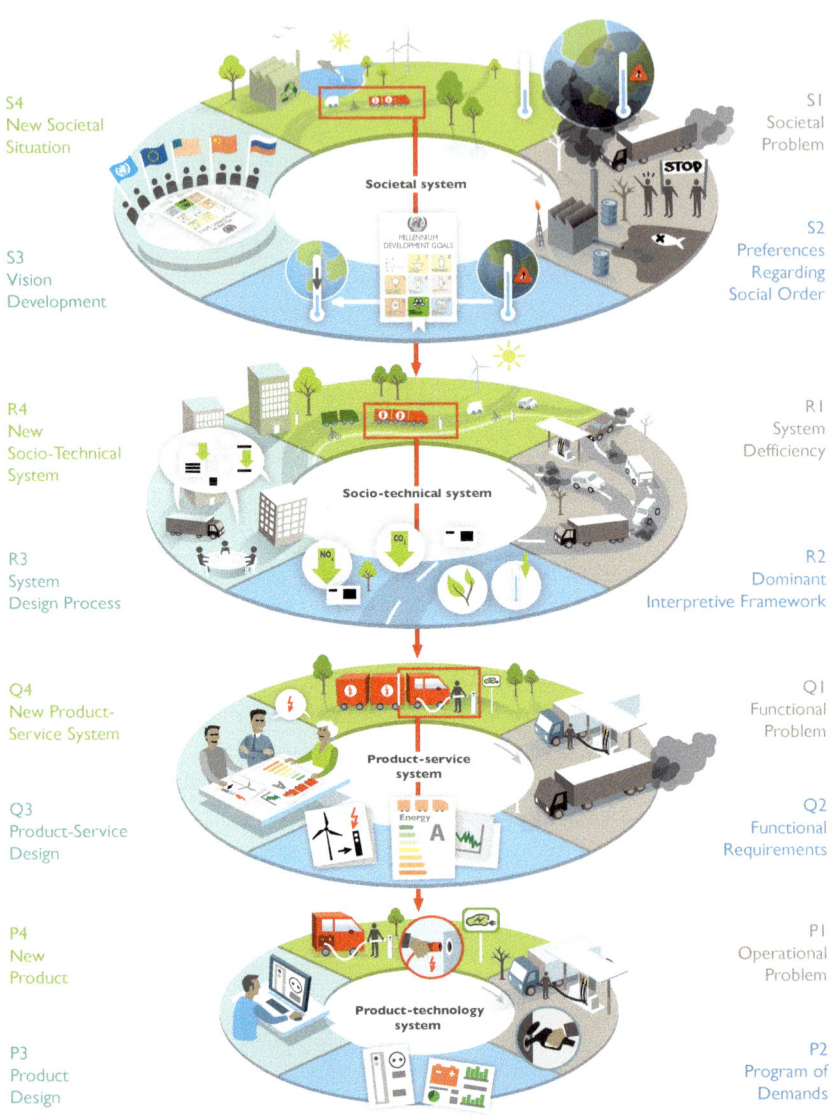

S4
New Societal
Situation

S3
Vision
Development

S1
Societal
Problem

S2
Preferences
Regarding
Social Order

Societal system

R4
New
Socio-Technical
System

R3
System
Design Process

R1
System
Defficiency

R2
Dominant
Interpretive Framework

Socio-technical system

Q4
New Product-
Service System

Q3
Product-Service
Design

Q1
Functional
Problem

Q2
Functional
Requirements

Product-service
system

P4
New
Product

P3
Product
Design

P1
Operational
Problem

P2
Program of
Demands

Product-technology
system

Figure 2
Multilevel Design Model [11]

Finally, we tested the system with the employees at Nike's European headquarters in Hilversum, with participants documenting their experiences in a diary. The new transport system was now used on a small scale, and the lessons we learned from it were to be translated into application on a large scale. This approach is also described as a Strategic Niche Experiment,[6] a Bounded Socio-Technical Experiment,[7] or a Transition Experiment.[8]

78

Over the years, researchers have worked on this methodology, and started to use different names for somewhat similar approaches.[9] Much of this research applies a more philosophical and sociological perspective; it is not aimed at designers. I tried to establish the connection with design in my dissertation, where I developed a Multilevel Design Model (Figure 2) to describe the relationship between the different system levels at which the design process takes place.[10] [11]

Innovation at the cutting edge of industries

This dissertation was the first step toward a position as a professor of Open Innovation at the NHL Stenden University of Applied Sciences in Leeuwarden. In that role, it was my task to link the various professional areas of the university of applied sciences with the idea that innovation takes place at the cutting edge of different fields of work. This is where the Neue Kombinationen (new combinations) are created, as Joseph Schumpeter already mentioned in 1911.[12]

This cross-sector approach is also essential in another ambition of the research group, which aims to develop solutions to the complex societal issues that we face. For example, the northern part of the Netherlands wants to lead the quest for circularity. This, however, requires more than, for example, technical solutions aimed at recycling plastic waste. It requires economic profit models, behavioral change, new policies and legislation, to name a few. In short, the ambition towards sustainability and circularity requires a multidisciplinary, interdisciplinary, or transdisciplinary systemic approach.

Everyone designs

The research group has been around for more than thirteen years now, and NHL Stenden has adopted the design process as a leading educational concept for the entire university. Under the name Design-Based Education, more than 20,000 students from more than 75 different courses daily work on developing new solutions for all different sectors of society.

6. René Kemp, Johan Schot & Remco Hoogma, "Regime Shifts to Sustainability Through Processes of Niche Formation: The Approach of Strategic Niche Management," *Technology Analysis & Strategic Management*, 10:2 (1998), 175–198, https://doi.org/10.1080/09537329808524310.

7. Halina Szejnwald Brown, Philip Vergragt, Ken Green, Luca Berchicci, "Learning for Sustainability Transition through Bounded Socio-technical Experiments in Personal Mobility," *Technology Analysis & Strategic Management*, 15:3 (2003), 291–315, https://doi.org/10.1080/09537320310001601496.

8. René Kemp, Suzanne van den Bosch, *Transitie-Experimenten – Praktijkexperimenten met de Potentie om bij te dragen aan Transities* (Delft: Kenniscentrum voor Duurzame Systeeminnovaties en Transities, 2006).

9. Frans Sengers, Anna J. Wieczorek, Rob Raven, "Experimenting for Sustainability Transitions: A Systematic Literature Review," *Technological Forecasting and Social Change* 145 (2019), 153–164.

10. Peter Joore, *New To Improve: The Mutual Influence Between New Products and Societal Change Processes*, (PhD dissertation, Delft University of Technology, 2010).

11. Peter Joore, Han Brezet, "A Multilevel Design Model – The Mutual Relationship Between Product-Service System Development and Societal Change Processes," *Journal of Cleaner Production* 97 (2015): 92–105, https://doi.org/10.1016/j.jclepro.2014.06.043.

12. Joseph Schumpeter, *The Theory of Economic Development* (Cambridge: Harvard University Press, 1911).

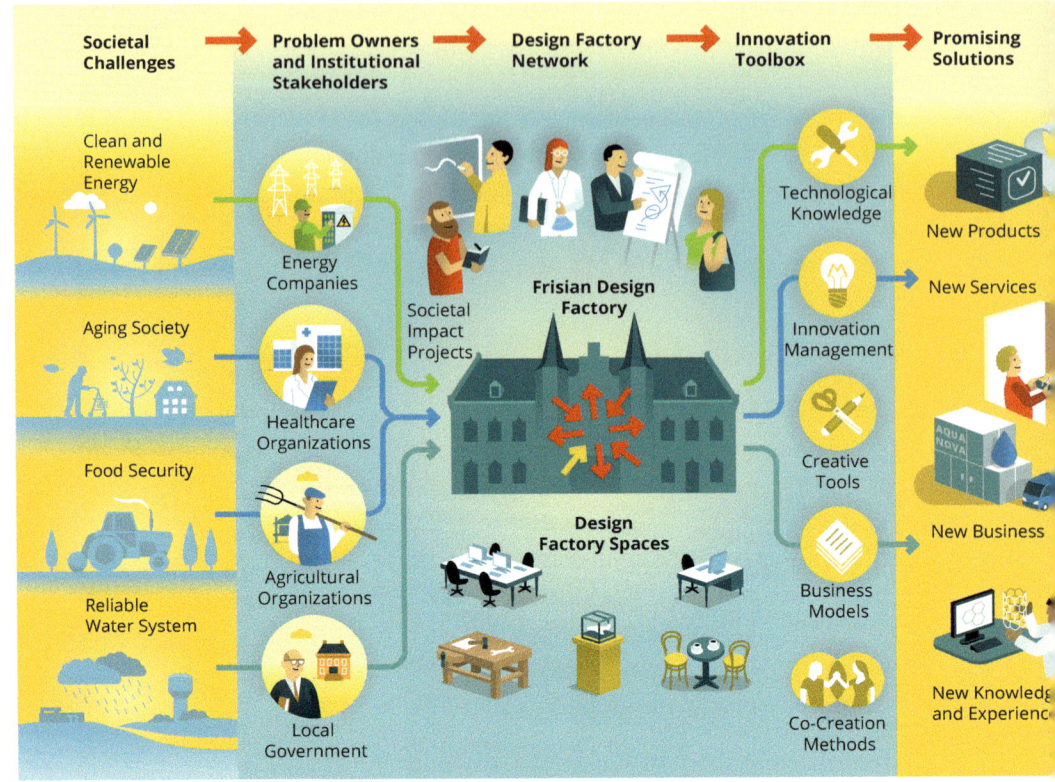

Figure 3
Visualisation of the systemic approach that is used by the Frisian Design Factory to design solutions for complex societal challenges. Illustration by Marc Kolle.

Currently, my university of applied sciences has dozens of design workshops where students design solutions for questions from professional practice. These solutions are certainly not always physical objects or products but can also be in the form of a game, a recommendation, a policy plan, a care protocol, or a business plan. In fact, it means that all professionals are considered designers. This was also described in 1969 by Nobel laureate Herbert Simon. He stated in his book *The Sciences of the Artificial* that '*Engineers are not the only professional designers. Everyone designs who devises courses of action aimed at changing existing situations into preferred ones. The intellectual activity that produces material artifacts is no different fundamentally from the one that prescribes remedies for a sick patient or the one that devises a new sales plan for a company or a social welfare for a state. Design, so construed, is the core of all professional training: it is the principal mark that distinguishes the professions from the sciences. Schools of engineering, as well as schools of architecture, business, education, law, and medicine, are all centrally concerned with the process of design.*' [13]

Designing a miniature society

We are now facing the same challenge as I described above in the pursuit of a sustainable society. Although most of the questions from professional practice can be answered with monodisciplinary solutions, a cross-sector perspective is necessary to address the real complex societal challenges. To achieve this, the concept of the strategic niche experiment mentioned above can be translated into a design environment. This could be described as a field lab or a living lab, where we would be working on groundbreaking solutions on a type of 'intermediate scale'. The working level here must be 'large' enough to think at the societal system level. And at the same time, it must be 'small' enough to make the solutions that have been developed concrete and tangible.

We try to apply this approach at the Frisian Design Factory located in the former Blokhuispoort Prison in Leeuwarden. Here, students, lecturers and professionals cooperate on solving complex challenges related to energy, water, food or healthcare, as presented in Figure 3. One of such examples in which we work at the level of the societal ecosystem is a collaboration with various stakeholders on the Frisian or Wadden Islands. Together with local authorities and entrepreneurs in the hospitality industry, students are working on the ambition to make the islands completely plastic-free. Innovations include the use of durable materials such as biodegradable plastics, but the project focuses even more on avoiding the use of plastics altogether. One of the ways we try to achieve this is by using 'nudging': designing the environment so that visitors are more or less seduced to display the desired behavior. We do this, for example, by making reusable products much more accessible compared to the less desirable disposable products.[14]

Another example of applying this systemic design approach is the Inno-Quarter project, where we develop sustainability solutions at and with festivals. For example, we created an environment called DORP (the Dutch word for village) at the Welcome to the Village festival. Here, students design solutions for sustainability issues. Because the entire festival is built from scratch in one week and is taken down afterward, a complete miniature society is built from scratch.

13. Herbert Simon, *The Sciences of the Artificial*, Third Edition (Cambridge, MA: MIT Press, 1996).

14. Marcel Crul, Plastic-Free Tourism and Hospitality on Dutch Wadden Islands: Multi-level Design Approaches and Experiences. *Proceedings of European Roundtable for Sustainable Consumption and Production* (Graz, 2021).

15. Aranka Dijkstra and Marije Boonstra, *Festival Experimentation Guide,* (Leeuwarden, NHL Stenden Publishers, 2021).

16. Aranka Dijkstra, Sybrith Tiekstra, Gertjan de Werk, Peter Joore, "Festivals as Living Labs for Sustainable Innovation: Experiences from the Interdisciplinary Innovation Programme DORP," *Proceedings of European Roundtable for Sustainable Consumption and Production* (Barcelona, 2019).

Something similar takes place at the Into the Woods festival in Sweden and the Northside festival in Denmark. We recently described our working method in the Festival Experimentation Guide (Figure 4).[15] This 326-page manual contains dozens of examples of innovations developed or tested at the festivals, ranging from the Semilla Sanitation Hub (which converts urine into drinking water), the Comp-A-Tent (a compostable tent based on bioplastics, hemp and cardboard), KlimaKarl (a CO_2 reduction game by a startup in Bremen) and SaruSoda (an organic post-mix lemonade). The challenge is still to really innovate at a systemic level, but the foundation has been laid.[16]

Figure 4
Festivals as a breeding ground for innovation, the Festival Experimentation Guide.[15]

Idealistic visions and realistic solutions

Finally, what about applied design research and all the related definitions? I have to say that design research has evolved considerably, when I look back at the bouncing vacuum cleaners I mentioned at the beginning of this article. Nowadays, we emphasize the difference between research 'for', 'into' and 'through' design, while also still using the difference between the 'designer' who wants to change the world, and the 'researcher' who wants to understand the world. These roles seem to be increasingly intertwined. After all, to change the world effectively, you must first understand it properly. And to understand the world properly, it is actually essential to work with it in a practical environment.

In this sense, I feel least connected to the 'pure' researcher, who studies the world remotely without actively entering the playing field. Although this is the most neutral and objective approach from a scientific point of view, it does not appeal to me as a designer. At the same time – but perhaps that has something to do with age – I find that in recent years, I have been using an increasingly more reflective perspective, trying to truly understand the innovation ecosystem. Perhaps with the idea that only if you understand a situation properly can you design effective interventions. Precisely that is what has been the consistent factor in the more than 35 years that I have been working in design. And that is what makes design and designers so interesting and relevant, as far as I am concerned. It is ultimately about developing a better and more beautiful world, in which idealistic visions of the future are translated into realistic solutions in the here and now. This seems to have finally bridged the apparent contradiction between realism and idealism that I once struggled with as an 18-year-old student.

Peter Joore

NHL Stenden University of Applied Sciences

Dr.ir. Peter Joore focuses specifically on design processes in which different types of actors, across the sectoral boundaries, work together to solve complex societal issues in a living lab environment. He was trained as an industrial designer at TU Delft, where he also obtained his PhD in 2010. After graduating in 1991, he worked as a designer at several companies. He started working at the Netherlands Organisation for Applied Scientific Research (TNO) in 1999. In 2008, he switched to higher education, working as a professor of Open Innovation at the NHL Stenden University of Applied Sciences in Leeuwarden.

Designing the future

Navigating the future

Tessa Cramer

As professor of Designing the Future, it is my mission to promote futures literacy. And in my work as a futurist, I connect the past and the future to help others make informed choices today. In my work, I not only build bridges between the past and the future, but also between science and design. In May 2020, I obtained my doctorate with my dissertation 'Becoming Futurists,' detailing how futurists can be understood as a profession. Writing this academic work made me realize that the mindset of futurists can be useful for others too. For example, futurists share a mindset in which creativity and keeping an open mind are regarded as valuable. And futurists are not impressed by disciplines or boundaries; they like to reside in the in-between space of not knowing and uncertainty. In my research group, I translate these lessons to a broader audience, for example, for people who do not have the privilege yet to think about the future, and invite them to learn new skills for navigating the future.

Together with an alternating team of researchers that includes lecturers and students, I strive to help others increase their knowledge about the future and raise new questions. We do this, for example, by questioning how to live with the consequences of the COVID-19 pandemic. Responding adequately to the changes that are currently taking place is quite challenging. Words such as fear, uncertainty, and change are used frequently and are often

1. Slavoj Zizek, *The Courage of Hopelessness, Chronicles of a Year of Acting Dangerously,* (London: Penguin Books, 2017).

2. Kate Raworth, *Doughnut Economics, Seven Ways to Think Like a 21st-Century Economist* (White River Junction, VT: Chelsea Green Publishing, 2017).

3. Roman Krznaric, *The Good Ancestor, How to Think Long Term in a Short Term World* (London: Penguin Books, 2020).

considered to be negative. But the future offers hope at the same time. There is plenty of room to research and provide solutions that offer added value from a social point of view.

The research question we ask at the Designing the Future research group against this background is: how can people learn to deal with uncertainty? In more concrete terms: which ideas or (online) products and services can be designed to learn how to deal with uncertainty? These questions are particularly relevant to the students of Fontys Academy for Creative Industries (from now on: Fontys ACI). They shape that future in all its diversity, for example, by developing new online business models in the Digital Business Concepts training program, creating new meaningful and sustainable experiences in International Event Music & Entertainment Studies, and designing meaningful lifestyle concepts in Trend Research & Concept Creation.

The relevance of a design approach to explore the future

Both on a methodical and philosophical level, a design approach to explore the future is relevant for a broader audience. From a philosophical perspective, in 2013, we were struck by the words of Ricardo Semler in the Dutch TV show *Tegenlicht*: "We have become boxed people." He describes how 'boxes' dictate our lives, not only at work, through flow charts, but also at home (in architecture) and on the road (in car or train). Semler describes how difficult it is to reflect on the system we are part of. The philosopher Zizek also describes this in his work the *Courage of Hopelessness*, with the firm statement that we are letting ourselves be distracted while Rome is burning.[1]

Another thinker who inspired us is Kate Raworth. In her work *Doughnut Economics*, she states that progress should not be achieved at the expense of our planet.[2] In the book *The Good Ancestor*, Roman Krznaric makes a plea for taking responsibility today to safeguard the well-being of future generations.[3] Each of these thinkers shows us that issues have become so complex that we need all of our creativity and imagination to even begin to navigate them.

Figure 1
Students at work at Fontys
Academy for Creative
Industries.

The Designing the Future research group is aware that future solutions require new, fresh perspectives. Old systems will not help formulate relevant answers. Young generations of students are not yet burdened with too much knowledge of these systems and as a consequence, are able to 'hack' them in a friendly way. Our research group sees it as its task to facilitate curiosity about the future, enabling students and lecturers to contribute to shaping the future in a focused way.

The connection between design thinking, future research, and speculative design

But how do we methodically translate these insights into a design approach? In the research group, we emphasize 'designing' in a broad sense. Note that we use the verb ('to design'). That is on purpose: my institution trains creative thinkers who develop innovative concepts that contribute to society.

There are three areas of methodical expertise that we develop within the research group: design thinking, speculative design, and futures research. *Design thinking* is an accessible and practical approach that is already used in many different contexts and training programs. The method

4. Tim Brown, *Change by Design. How Design Thinking Transforms Organizations and Inspires Innovation* (New York: Harper Collins Publishers, 2009).

5. Roger Martin, *The Design of Business. Why Design Thinking is the Next Competitive Advantage* (Boston, MA: Harvard Business Press, 2009).

6. See for example Jeanne Liedtka, "Why Design Thinking Works," *Harvard Business Review* (September October 2018) and Michael Lewrick, Patrick Link, Larry Leifer, *The Design Thinking Playbook: Mindful Digital Transformation of Teams, Products, Services, Businesses and Ecosystems* (New York: Wiley, 2018).

7. Antony Dunne, Fioana Raby, *Speculative Everything. Design, Fiction, and Social Dreaming* (Cambridge, MA: MIT Press, 2013).

8. James Auger, *Speculative Design: Crafting the Speculation*. Digital Creativity 24, no. 1 (2013): 11–35, https://doi.org/10.10 80/14626268.2013.767276.

9. For more information about the Artificial Womb, see the NextNature website at https://next-nature.net/story/2018/artificial-womb-design.

offers our students concrete steps that they can apply in the research phase to develop concepts. Design thinking has long existed within the domain of designers but was mainly translated into new contexts and worlds by publications and works by Tim Brown [4] and Roger Martin.[5] In recent years, this method has gained enormous ground, both academically and in practice. In addition to the extensive literature published on the subject,[6] many large companies such as IBM, Samsung, and Philips make use of (elements of) design thinking.

Speculative design is less well-known and widespread than design thinking but appeals to the imagination. This perspective invites designers and researchers to make the future tangible. In 2013, Dunne and Raby published the book *Speculative Everything,*[7] which can be seen as the starting point for speculative design. The central thesis of speculative design is that ethical discussion about what do we want with our future, is sparked when that future is tangible. The value of speculative design is, on the one hand, that this approach encourages reflection on possible futures and, on the other, to learn to ask critical questions.[8] Speculative design can take many forms; it can be a science fiction film, a tangible object, or a poem. For example, NextNatureNetwork [9] developed an artificial womb, which raises the question: is the development of an artificial womb desirable?

Finally, *futures research* is often connected to methods like scenario planning. With this perspective it is possible to think in alternative futures and question what one cannot know about the future.[10] This requires a very different mindset than focusing on what we do know about the future.[11] The latter approach, focusing on the known, is tricky because it encourages us to think about the future in a linear way. Today is the same as tomorrow, even if we know that this is not the case. Uncertainty is a central concept in this perspective [12]; the future is simply impossible to predict, and the courage to embrace that we do not know is of vital importance. This is also why futurists, artists, and designers have a lot in common: they all navigate, in their own respective ways, that gray area of not-knowing.

Within the research group, we see the added value of connecting the various perspectives: *future research, design thinking, and speculative design*. We make the future tangible at multiple levels by connecting the design thinking approach to the creation of tangible concepts with speculative design. It makes the somewhat abstract perspective of futures research concrete and adds an additional layer to the further development and deepening of design thinking and speculative design. We do this, for example, by making it clear how to create your own desirable future guided by the three perspectives. We also encourage our researchers to experiment with the perspectives and fail, so they can try again. We believe that this approach is useful to grapple with the ever-evolving and changing world around us.

Highlighted project: the Rose Garden

The global health crisis caused by COVID-19 was an invitation to slow down under exceptional circumstances: we found ourselves collectively in the eye of an unexpected storm. Within the Designing the Future research group, we noticed an increased understanding of what matters, both socially and personally. We momentarily let go of the focus on theory and methods, so we could openly observe what was going on around us. Our colleagues and students started to raise new questions. For example, what is 'essential' when the economy is on hold? I often overheard colleagues ask: Why was I in a hurry again? Together we started to question: Why is an attentive, slower paced environment the exception rather than the rule?

There was no straightforward answer to the questions that were raised. They take time and require patience. This also had practical reasons; it proved to be difficult to create headspace if long to-do lists need to be finished. We started to narrow our search down to the question of how to escape the mechanisms in our society that consider our attention as a currency, vying for our attention, visibly and invisibly. Our era is characterized by a high degree of acceleration, in which time seems to be passing increasingly faster. We use multiple screens simultaneously, constantly engulfed by notifications and apps.

10. E.g., Wendell Bell, *Foundations of Futures Studies: Human Science for a New Era, Volume 1* (New Brunswick: Transaction Publishers, 2003).

11. Tessa Cramer, *Becoming Futurists: Reluctant Professionals Searching for Common Ground*, PhD Thesis (Maastricht: Maastricht University, 2020). https://doi.org/10.26481/dis.20200520tc.

12. For a more in-depth view of dealing with scenarios and uncertainty in practice, I refer to Marjolein van Asselt et al, *Foresight in Action, Developing Policy-Oriented Scenarios* (London: Earthscan, 2010).

Figure 2
Studio of the Designing
the Future research group
in the Fontys Academy for
Creative Industries.

We then went looking for literature and practices that could help us in the search for answers. We found reflective books, such as publications by Ramsey Nasr [13] and Merlijn Twaalfhoven,[14] and a wide range of academic publications that can be considered future research.[15] [16] [17] Our search was triggered by several officials asking us: How do we navigate this uncertainty since we've always been guided by what is certain? That requires a tremendous change in thinking and doing. Through her book *How To Do Nothing: Resisting the Attention Economy*, artist and writer Jenny Odell provides us with a valuable metaphor to envisage that headspace: the Rose Garden.[18] Time and again, she returns to a rose garden in her neighborhood to experience that she doesn't have to do anything. The rose garden showed her, and us, that proximity to other people is not always necessary to connect. And sometimes silence can connect more than breaking it. In a sense, this rose garden resides in all of us, and we can make it work for us – provided we make the time for it. The latter is not always easy, with a smartphone at your fingertips.

From this starting point, I started working in a special task force, set up just after the first lockdown in April 2020, which lasted until July 2020. The experimental task force, consisting

90

of secretaries, translators, and lecturers, was highly agile. For example, graduating students received mental support from the taskforce with a mobile fair that went door to door to visit them. This action was soon picked up by the national press and was reported in Dutch newspapers *Telegraaf* and *Algemeen Dagblad*. Another example was a large group of lecturers who went on a search for headspace, at home, online, and in the curriculum. Hiking clubs were started, as well as new peer review groups. All this work laid the foundation for a larger theoretical project that resulted from the initial question: How can people learn to deal with uncertainty? The project that resulted from this successful task force was renamed the Rose Garden.

In the current academic year (2020/2021), the research group is looking for ways to make the Rose Garden a place for students, lecturers, Fontys ACI staff, and Fontys-wide employees, but also for externals. The first concrete attempt is to facilitate a Floating Festival about 'nothing'; during a period of ten weeks, this festival 'floats' online through the corridors of Fontys ACI. From May to the summer holidays in 2021, every day from noon until 1 pm, we hosted an open mic for all employees and students who needed company and to just be together. We were surprised by how creative participants became: we saw them recite poems, play piano, philosophize, and share research results, play games, write together and display artwork. The festival allowed our colleagues and students to slow down, stop doing things and just be present. We hope to expand the project next year, designing more physical and digital Rose Gardens in collaboration with others.

The societal value of the design approach

The challenge for the coming years is to let the forward-looking design approach also inspire other disciplines and experts, such as politicians. Last year, for example, we were inspired by Ramsey Nasr's interview on the Dutch TV show *Buitenhof* about the value of creatives. In April 2020, he noted that in the midst of the health crisis, the main focus

13. Ramsey Nasr, *De Fundamenten* (Amsterdam: De Bezige Bij, 2021).

14. Merlijn Twaalfhoven, *Het Is Aan Ons. Waarom We de Kunstenaar in Onszelf Nodig Hebben om de Wereld te Redden* (Amsterdam/Antwerpen: Atlas Contact, 2020).

15. Krista Tippett, *Becoming Wise. An Inquiry Into the Mystery of Art of Living* (Penguin Press, 2016).

16. Robin Wall Kimmerer, *Braiding Sweetgrass. Indigenous Wisdom, Scientific Knowledge and the Teaching of Plants* (Milkweed Editions, 2015).

17. Louise Byg Kongsholm and Cathrine Gro Frederiksen, *Trend Sociology V.2.0 – the Ultimate Guide: Theoretical, Methodical and Practical Work With Trends* (Pej Gruppen, 2018).

18. Jenny Odell, *How To Do Nothing, Resisting the Attention Economy* (Brooklyn, NY: Melville House, 2019).

91

Figure 3
Future thinking for
beginners, a drawing made
during the Chaos in Order
festival in November 2020.
Credits: Joni Israeli.

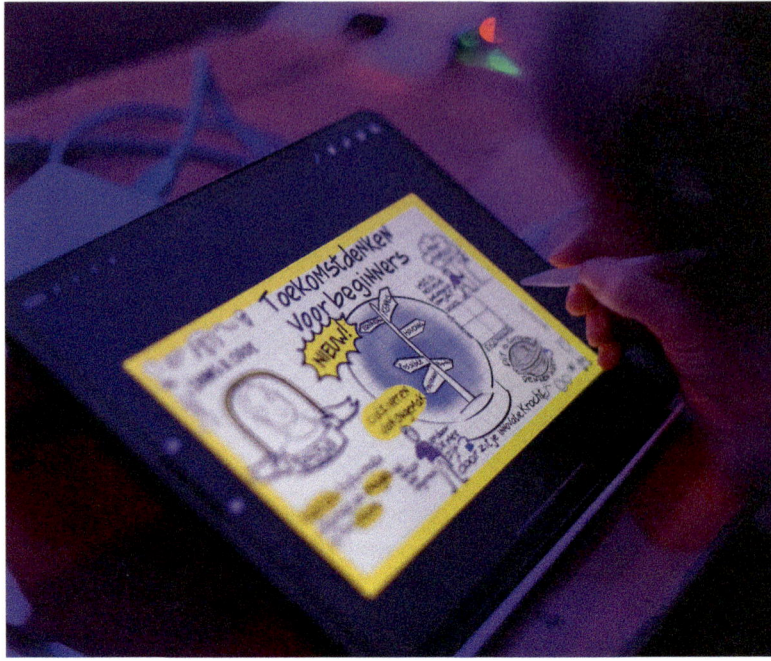

was on 'valid' knowledge developed by a specific type of
scientist. The creatives, artists, and designers could have
made a massive contribution creatively thinking about the
social distancing guidelines. However, that connection has
not been structurally established. The one-sided image of
creativity as entertainment is outdated, but apparently still
very persistent. It is time to expand the societal narrative on
creativity and the creative economy and show their value
as sources of unexpected inspiration to inform decision-
making at the local and national level. Partnerships such as
the NADR can contribute to this broader understanding and
appreciation of the design disciplines.

Tessa Cramer

Fontys University of Applied Sciences

Dr. Tessa Cramer is futurist and professor of Designing the Future at the Academy for Creative Industries at the Fontys University of Applied Sciences. She is a cultural sociologist and obtained her PhD from Maastricht University in 2020 with a dissertation titled 'Becoming Futurists.' Cramer is co-founder and former board member of the Dutch Future Society, co-founder of the Trend Research Lab, created a bachelor curriculum about the future, is a member of the New Amsterdam Council of Pakhuis De Zwijger as well as a member of the Development Council of BrabantKennis and a public speaker via TheNextSpeaker. In her work, Cramer combines design and future thinking by applying design principles to develop solutions for the future.

The artistic attitude in a social context

The art of daring to see things differently

Anke Coumans

The research group *Image in Context* of the Research Centre *Art and Society* in Groningen has spent the past eight years developing new roles for artists and designers working in social contexts. While searching for the specific role that the arts (as a collective term for the various disciplines offered within an art academy) can play in social environments, we discovered a particular artistic attitude by which an artist is distinguished beyond their visual and image-making quali- ties. After all, without this artistic attitude that is the baseline for all artistic practice, visual qualities would lose their full potential, and the artistry would become but craft. It is this artistic attitude that, in a social context, marks the difference between the artist and, for example, a social worker or a healthcare professional. The artistic attitude offers new and radical avenues to look at social contexts beyond the care context itself. This attitude is developed during their time at the art academy.

95

1. Merlijn Twaalfhoven, *Het is Aan Ons. Waarom We de kunstenaar in Onszelf Nodig Hebben om de Wereld te Redden* (Amsterdam/Antwerpen: Atlas Contact, 2020).

2. Twaalfhoven, *Het is aan Ons*, 21.

3. Elizabeth Fisher and Rebecca Fortnum, eds., *On Not Knowing: How Artists Think* (London: Black Dog Publishing, 2013).

4. Herman van Hoogdalem and Gijs Wanders, *Gezichten van Dementie* (Zwolle: WBOOKS, 2016) and: Herman van Hoogdalem and Constance de Vries, *Mag Ik Gaan. Leven en Sterven met Dementie* (Zwolle: WBOOKS, 2020).

Few publications have been written about this artistic attitude, and even less within the academic context. The only exception is composer Merlijn Twaalfhoven's 2020 book *Het is aan ons. Waarom we de kunstenaar in onszelf nodig hebben om de wereld te redden* ('It is up to us: Why we need our inner artist to save the world'),[1] in which he examines the artistic attitude through his own professional experience. Twaalfhoven speaks about the mentality of the artist in this context. To help non-artists take ownership of this type of attitude too, he offers a tapestry of observations, experiences, thoughts and actions as an instruction for how to look and see differently, dare to feel more, learn to think expansively and start advancing to reach one's ideals.[2] In his book, he explores these four artistic mentalities through successful and less successful attempts to turn his audience into participants.

In this contribution, I will highlight how the focus on the artistic attitude has developed using one of the research projects of the research group: *Ik zie ik zie wat jij niet ziet. Portretten van mensen met dementie* ('I spy with my little eye: Portraits of people with dementia'). Using various artist's texts from the book *On Not Knowing: How Artists Think* [3] and the thinking of anthropologist Tim Ingold, I will explore the very nature of this artistic attitude. Finally, via a second research project of my research group entitled *De ontwerpende attitude in de zorg voor mensen met dementie* ('The design-like attitude towards people with dementia in care institutes'), I will clarify the central focus of my research group for the coming five years. By researching how the artistic attitude can be adopted by those who did not study at an art academy, I will focus primarily on professionals working in various care practices.

A focus on the artistic attitude

For the project *Ik zie ik zie wat jij niet ziet,* we asked art students to portray people with dementia. This project was developed with my colleague, visual artist Herman van Hoogdalem, whose portraits of those with dementia have been central to his own practice over the past decade.[4] Van

Hoogdalem was the experiential expert in this research, engaging with people with dementia via the process of creating these portraits.

In the first two years, 2015 and 2016, my research group ran this project in the institutional context of a care home, where the art students and those with dementia were paired off. From 2017, the students were connected to the Odensehuis in Groningen: a drop-in center for those with dementia and their family members. With nothing more than a drawing pad, the very device that legitimized their presence in this context, the art students entered, with hesitation and without tangible points of reference, this world of dementia. They had some vague idea of what awaited them, often due to personal experience. Still, none of them

Figure 1
Drawing of the conversation we had with the informal carers, caregivers and the management of 't Blauwbörgje. Drawing and photo: Asa Scholma.

knew how these men or women would truly react to their presence. Alongside the finished portraits, which often were powerful testimonies of the meetings between students and people with dementia, the care professionals particularly noticed the characteristic manner in which the art students were present. They talked about the gentle slowness of a process wherein things might be noticed that may normally

97

fall through the cracks of care professionals view, because their focus lies more on quick and immediate tasks and 'helping.' The students were also warmly welcomed within the Odensehuis because of their capacity to see and listen to the people there and to make contact with them via the portrait-making process.

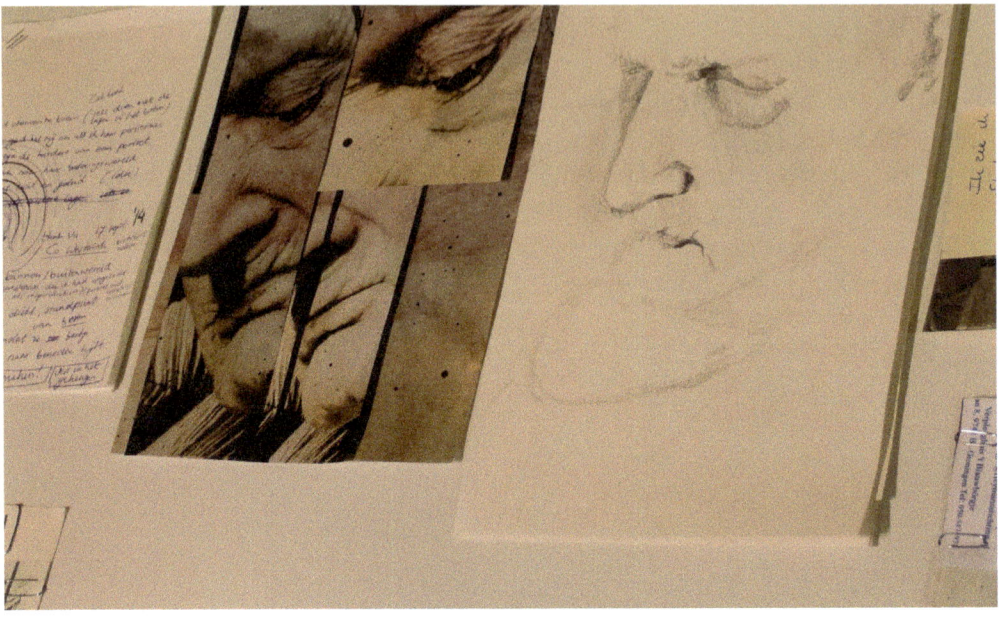

Figure 2
Works set-up for Asa Scholma's drawing process during the 'Ik zie, ik zie wat jij niet ziet' project. Photo: Asa Scholma.

The nature of the artistic attitude

In her article 'Tactics for not knowing: preparing for the unexpected', [5] Emma Cocker describes the artistic attitude of not-knowing when entering an unfamiliar reality. The artist gives space for the unexpected to occur, precisely through this characteristic of not-knowing. Already from a young age, we learn to identify and classify the unknown. Our education teaches us that we must expand knowledge, and that not-knowing is a fault. It seems that within the artistic professions, this not-knowing is, however, a necessary attitude that should somehow be protected.

In the 2013 Fortnum article, 'Creative Accounting: not knowing in talking and making', artist Terry Diffey is cited on this not-knowing, namely the absence of predictability in the creative process. "To create is to engage in undertakings the outcome of which cannot be known or defined or predicted, though there may be some presentiment of the outcome." [6] So, there is indeed a sense of direction present, but no predeterminate idea about the outcome.

According to anthropologist Tim Ingold, this artistic attitude potentially makes artists better anthropologists. In his 2017 Groningen lecture, Ingold mentions four particular qualities that are the foundation for this argument.[7] The first quality is generosity. Artists are generous when they are not greedy in respect to the world around them; when they understand that we solely exist because of what others have given us; when they can receive what is offered with an open attitude, and when they give back that which does not belong to them. The second quality of the anthropologist-like artist is their ability to work not solution-oriented but open-ended.

Thirdly, the anthropologist-artist should be comparative. This means that they recognize that one overall perspective is not the only possibility. With this, the artist questions both their own approach in relation to the other person and acknowledges that the approach of another is also of value. Finally, the artist is *critical*: they do not accept the world as is.

The aforementioned idea of not-knowing is also related to these four qualities. They facilitate an open and dialogical relationship with their surroundings. In my article 'Relational drawing: The artist as anthropologist',[8] I describe how these characteristics are present amongst the art students in the *Ik zie ik zie wat jij niet ziet* project.

Planting the seed for further investigation

Drawing from the same attitude of the *Ik zie ik zie wat jij niet ziet* students, we went on to carry out further research projects within the health and care context. The openness of this responsive type of environment was what catalyzed the development of these projects. My colleague Ingrid Schuffelers and I call this the *process-orientated* approach of working.[9] It always starts with not-knowing, being open to the unexpected and working open-ended rather than solution-based; that is how we bring this framework to each

5.   Emma Cocker, "Tactics for Not Knowing: Preparing for the Unexpected," in: Elizabeth Fisher and Rebecca Fortnum, eds., *On Not Knowing: How Artists Think* (London: Black Dog Publishing, 2013).

6.   Rebecca Fortnum, "Creative Accounting: Not Knowing in Talking and Making," in: Elizabeth Fisher and Rebecca Fortnum, eds., *On Not Knowing: How Artists Think* (London: Black Dog Publishing, 2013).

7.   Tim Ingold, *Art, Science and the Meaning of Research*. Keynote lecture presented at the symposium Thought Things (Groningen, November 2017).

8.   Anke Coumans, "Relational Drawing. De Kunstenaar als Antropoloog," *FORUM+ voor Onderzoek en Kunsten* 26, no.1 (2019): 38–47.

9.   Anke Coumans and Ingrid Schuffelaars, "De Relevantie van Artistiek Onderzoek," *ScienceGuide*, 21 June 2017.

of the interventions. We use this to process in the projects to open possibilities for how we conduct our researches. We developed these research projects in collaboration with the care workers, which resulted in a specific atmosphere and space where those involved could establish their own way of speaking, acting and being.

Figure 3
Online meeting while Willemijn Rog makes a drawing of the scenario outlined by the participants. Photo: Asa Scholma.

In this manner, we developed a methodology via the project *De ontwerpende attitude in de zorg voor mensen met dementie* whereby both the healthcare professionals and the family members of those with dementia (in this case, within the institutional care home setting of 't Blauwbörgje) learn from the different roles and perspectives present and gain a better understanding. We gave space to an attitude we called *artistic*. The people developed a similar open, receptive attitude towards their own surroundings as an artist would. We also called this attitude *design-like* in that it aims to actually change and improve the world that the care professionals and family members are a part of. This project was developed from my own opinion of care institutions, recognizing that they are designed around what is best for

100

the specific people with dementia, while the institutions should also cater to the actual needs of the healthcare workers and the family members.

This way of working closely touches upon Participatory Action Research as we very much consider the participants as co-researchers of their own context. I describe the framework of this research in the article 'Design in the here and now: The artistic attitude within the care context for people with dementia'.[10]

In 2020, we used this methodology with the same group of people from 't Blauwbörgje to gain insights into how the pandemic measures affected the inner workings, the nuts and bolts, of the care facility. The research focused on the disrupted relationship between those directly involved with 't Blauwbörgje and the outside world, particularly the relationship with young people. The development of a physical meeting space for the young and the elderly, for those outside and within 't Blauwbörgje, is the objective: it gives us the direction we wish to take with this project, once the situation allows for it.

Figure 4
Guyonne van Berge Henegouwen with portret of visitor of the Odensehuis in Groningen.

10. Anke Coumans, "Ontwerpen In Het Hier en Nu. De Artistieke Attitude In de Zorg Voor Mensen Met Dementie," *FORUM+ voor Onderzoek en Kunsten* 27, no.2 (2020): 3–13.

11. https://www.toukomst.nl/projecten/woonvormen/

12. Toukomst is a Groningen National Program initiative comprising of West 8 (urban development and landscape architecture design bureau) and other organizations from the province of Groningen. Together with Toukomst we aim to develop projects for the future of Groningen.

Research aims of the research group for the upcoming five years

The future challenges stemming from the projects described are fourfold. Firstly, we wish to collaboratively develop a new and actual place for people with dementia, their families, healthcare professionals and artists to live and work together. Our methodology can help us to form a growing development and design team. To achieve this, we have become part of the *Toukomst* [11] initiative's project group 'How we wish to live'.[12]

Secondly, we wish to research how the artistic attitude can help us to develop a health care system with humanity at its core. How can the residents, their family members and the care professionals develop their own practices within the overall policy framework? The pandemic has shown us how important it is to have policies based on the reality of what happens in the actual care context.

Thirdly, we are looking for a way in which this new way of working of artists and designers – designing environments in which the other person can become a designer – can be taught within art education. What is needed to train artists and designers to become designers of settings in a care context? What competences do they need for this?

Fourthly, we are researching whether or not the artistic attitude can, in fact, become a component of the care professional's education program. Just as the development of an artist can benefit from discovering a balance in between both an artistic and social attitude, we want care professionals to unearth what it means to operate with an artistic attitude within the care context.

To realize these challenges, we have built a consortium in which the healthcare sector and artists are both represented. Evelyn Finnema, Professor of *Nursing Science and Education,* (representing the healthcare industry) and the care institutes 't Blauwbörgje (Dignis Groningen) and Sunenz (Haren) are willing to run and develop pilots within their organizations. In addition, the alumni of the *Ik zie ik zie wat jij niet ziet* projects will participate as representatives from the artistic field.

Anke Coumans
Hanze University of Applied Sciences

Dr. Anke Coumans has gained work experience in a broad field of art and culture, in the Netherlands and abroad. Her work has focused on design, film, media, visual arts and journalism. In 2010, she obtained her PhD from Leiden University with the research 'Als een beeld 'ik' zegt ... het dialogische betekenisvormingsproces van het publieke beeld'. After a career as a professor and research coordinator in art education (HKU Utrecht), she was appointed as a professor at the Art and Society Center of the Hanze University of Applied Sciences in Groningen in 2013. Since her appointment, she has been investigating the role of dialogue in collaborative art and design projects.

Looking for trouble

Raising and tackling problems through design research

Eke Rebergen, Sebastian Olma, Wander Eikelboom

Designers work on various products, campaigns, and platforms that significantly impact our daily lives. The products and services they design – literally shaping everyday life – reflect social prejudices, ideologies, and power structures. Increasingly, these aspects are critically examined in the most common objects and services, such as online platforms and shops,[1][2] or in the increasing number of products that operate based on AI and algorithms.[3][4][5] The design of passports [6] or that of something as simple as a coke bottle has also received critical attention.[7]

The design of these products cannot be disconnected from the ethical, political, and far-reaching social consequences. A maker who designs under the naive belief that to be "neutral" forgets the social complexities and ideological undercurrents that are irrevocably attached to each design, both to the design process and its result. That is why it is a fundamental competence of the maker to critically understand and challenge the ethical, political, social context of her work. This is a central aspect of the research that takes place at Avans' CARADT (Center of Applied Research for Art, Design, and Technology).

1. James Bridle, *New Dark Age* (London: Verso, 2018).

2. Dominic Pettman, *Infinite Distraction* (Cambridge: Polity Press, 2016).

3. Ivana Bartoletti, *An Artificial Revolution: On Power, Politics and AI* (London: The Indigo Press, 2020).

4. Adam Greenfield, *Radical Technologies. The Design of Everyday Life* (London: Verso, 2017).

5. Safiya Umoja Noble, *Algorithms of Oppression* (New York: NYU Press, 2018).

6. Mahmoud Keshavarz, *The Design Politics of the Passport* (London: Bloomsbury, 2019).

7. Laurent de Sutter, *Narcocapitalism* (Cambridge: Polity Press, 2018).

8. Robert Hewison, *Cultural Capital* (London: Verso, 2014).

9. Angela McRobby, *Be Creative: Making a Living in the New Culture Industries* (Cambridge: Polity Press, 2016).

10. Evgeny Morozov, *To Save Everything, Click Here* (New York: PublicAffairs, 2013).

11. Oli Mould, *Against Creativity* (London: Verso, 2018).

12. Justin O'Connor, "The Great Deflation. Arts and Culture After the Creative Industries," *Making & Breaking* 2 (2021).

13. Anand Giridharadas, *Winners Take All: The Elite Charade of Changing the World* (New York: Knopf, 2018).

14. Sebastian Olma, *In Defence of Serendipity. For a Radical Politics of Innovation* (London: Repeater Books, 2016).

15. Elizabeth Resnick, *The Social Design Reader* (London: Bloomsbury, 2019).

16. Bruce Nussbaum, *Is Humanitarian Design the New Imperialism?*, 7 June 2010, https://www.fastcompany.com/1661859/is-humanitarian-design-the-new-imperialism.

17. Tim Seitz, *Design Thinking and the New Spirit of Capitalism. Sociological Reflections on Innovation Culture* (Cham: Palgrave Pivot, 2020).

Superficial innovations and the creative industry

We believe that design must always be critical and must question ideology. This also means that designers need to engage with the current discourse surrounding the 'creative industry'. Today, this is particularly urgent as policy terms such as 'innovation' or 'value creation' largely determine the development of creative work (as propagated by the 'Topconsortium voor Kennis en Innovatie van de Creatieve Industrie' and similar organisations). For years, critics have bemoaned the dominance of commercial concepts of creativity that have led to the adaptation of models whose novelty consists in finding new ways to stimulate unnecessary consumption, creating hollow innovation rhetoric and exacerbating social inequity.[8] [9] [10] [11] [12] Following some of these critical approaches, CARADT, the research group Cultural and Creative Industries investigates how artists and designers can help shape a desirable future without falling prey to what the American journalist Anand Giridharadas [13] calls the ideology of 'MarketWorld': the idea that cosmetic design interventions can correct the neo-liberal destruction of recent decades in the areas of mental health, social welfare, democratic governance and the environment. We understand that this is, in essence, a futile exercise in 'changeless change',[14] which we would like to steer clear of.

More critical design traditions

Within the design field, there is a current that engages with more urgent design challenges and tries to find effective ways to address them. Often, this falls under the label of 'social design',[15] i.e., projects addressing social challenges and societal problems. However, the present-day application of such 'social' design projects often avoids the necessary analysis of the broader social context in which a specific local "problem" is "solved" by a design intervention. Thus, it tends to ignore the conflicts that are inherent to fundamental change. As a result of such lack of analytical rigor, even socially-oriented 'humanitarian design' projects have been exposed as a new kind of imperialism.[16] Social change through 'design thinking' has been described as a lucrative kind of 'business consulting' [17] that does nothing but confirm the status quo.[18]

Fortunately, a more stringent critical movement in the design field seems increasingly gaining ground in various 'critical' or 'speculative' design practices, explicitly seeking room for ideological dissent and critical experiment. [19] These types of projects don't focus on effective solutions or lucrative improvement but on addressing structural problems and preventing undesirable developments. Yet, we should not be satisfied with a mere call for 'free debate' or an emphasis on 'critical thinking' for a small privileged group of designers, which is always a risk when it comes to these types of 'discursive' projects.[20]

A specific type of applied design research

Hence, we wholeheartedly agree with designers and design theorists who recently began to address potential 'design struggles' within the creative fields,[21] and who want to radicalize the (training to become a) designer.[22] To do this, complex relationships such as that between design and colonialism,[23] or the role of creatives in violence,[24] must be explored and deepened. More attention must be paid to the long history of (often less well-known) designers and makers who tried to shape counter-movements or fundamentally questioned the status quo.[25]

In the design domain, efforts will have to be made to break the dominant paradigms and complicit narratives.[26] It should be obvious that this requires a better integration between design theory *and* design practice, academic research *and* technological experiments, critical reflection *and* creative action. In short, this is what we believe a timely integrated approach to applied design research requires. It also means that we need to change the scope of the label 'applied design research' to include an explicit challenge to the neoliberal ideology of creativity and innovation – much more than has been the case so far in either education or the field of design research.

18. Anand Giridharadas, *Winners Take All: The Elite Charade of Changing the World* (New York: Knopf, 2018).

19. Anthony Dunne and Fiona Raby, *Speculative Everything; Design, Fiction and Social Dreaming* (Cambridge: MIT Press, 2013).

20. Bruce M. Tharp and Stephanie M.Tharp, *Discursive Design. Critical, Speculative, and Alternative Things* (Cambridge: The MIT Press, 2019).

21. Claudia Mareis and Nina Paim, *Design Struggles. Intersecting Histories, Pedagogies, and Perspectives* (Amsterdam: Valiz, 2021).

22. Danah Abdulla, *Modes of Criticism 4. Radical Pedagogy* (Eindhoven: Onomatopee, 2019).

23. Tristan Schultz, Danah Abdulla, Ahmed Ansari, Ece Canlı, Mahmoud Keshavarz, Matthew Kiem, Luiza Prado de O. Martins and Pedro J.S. Vieira de Oliveira (2018) Editors' Introduction, Design and Culture, 10:1, 1–6, DOI: 10.1080/17547075. 2018.1434367.

24. Paola Antonelli and Jamer Hunt, *Design and Violence* (New York: The Museum of Modern Art, 2015).

25. Marjanne van Helvert, *The Responsible Object. A History of Design Ideology for the Future* (Amsterdam: Valiz, 2016).

26. Daniela K. Rosner, *Critical Fabulations. Reworking the Methods and Margins of Design* (Cambridge: The MIT Press, 2018).

107

Finding and stirring up problems

Some years ago, Marie L. J. Søndergaard [27] used an expression that we believe helps to characterize what applied design research should be all about. She obtained her PhD by developing a design approach that she described as 'Staying with the Trouble through Design'. This was inspired by the work of Donna Haraway (though it stayed clear of Haraway's complex concepts such as 'chthulucene' or 'cyborgs').

Figure 1
Screenshot of www.creatieveweerbarstigheid.nl.

Søndergaard's design approach is characterized by her unfaltering insistence on a specific problem (staying with the trouble) and, above all, by her making problems (creating trouble), rather than submitting to the design dogma of finding solutions for problems. By engaging with the world in such a meaningful way, general social problems and criticism did not remain abstract theoretical reflection. Instead, they turned into concrete phenomena that designers have to struggle with and are part of. Problems are thus brought in, raised, stirred up, and learned to be dealt with. This approach can be used in the present context to make a difference between, on the one hand, (common) forms of applied design research, where problems are the starting point for effective improvement or innovation, and on the

other hand, the new kind of design research that we consider to be so important, where designers learn to stay with the trouble.

At Avans University of Applied Sciences, we have used two different ways to try and shape this type of applied design research in educational projects.

Example of educational project 'playful interaction'

In a second-year quarterly project called 'playful interaction', two cohorts of more than 120 students of the CMD (Communication and Multimedia Design) program in 's Hertogenbosch have worked on playful interventions for social problems. This project uses the potential of games and playful interactions to (temporarily) create or challenge a different reality, or to disrupt and break through the usual patterns of the design process [28]. Students were divided into design teams and asked to come up with three design proposals. To inspire them, a collection of sample projects was gathered on the website www.creativeweerbarstigheid. nl (Figure 1), based on previously published inventories of this type of creative work.[29] Some of the student's research findings were added to the collection at the end of the project; next year, they can serve as examples for a new design team.

One of the student projects was a design for a new program called 'Medicinal Distribution and Management Analysis' (Figure 2), a draft proposal based on research into the problems of drug trafficking, created by students Kevin Nas, Bram Smits, Pleun Wilting, Luna van Loon, Mannus van der Meer and Damon van der Voort. The students created various communication materials, drafted an initial set-up of entry requirements and a draft program description. They continued the logical line of thinking about further professionalization and the significant economic importance of mostly illegal drug trafficking and wondered whether, and how, an extension of the training portfolio at their own university of applied sciences could be aligned to that.

27. Marie L. J. Søndergaard, "Staying With the Trouble Through Design: Critical-feminist Design of Intimate Technology," PhD Thesis (Aarhus University, 3 December 2018).v

28. Ian Bogost, Play Anything (New York: Basic Books, 2016).

29. Garnet Hertz, Disobedient Electronics. Protest, January 2018, http://www.disobedientelectronics.com/; Finn Brunton & Helen Nissenbaum, Obfuscation. A User's Guide for Privacy and Protest (Cambridge: The MIT Press, 2012); Nato Thompson and Gregory Sholette, The Interventionists, Users' Manual for the Creative Disruption of Everyday Life (Cambridge: MIT Press, 2004).

MEDICINAL DISTRIBUTIO
MANAGEMENT ANALYTI

VOLTIJD 'S-HERTOGENBOSCH

De grote wereld van de drugs is waar jij je thuis voelt. Je bent analyserend en ontwikkelend. Je werkt bij Medicinal Distribution & Management Analytics aan verschillende werkvlakken van deze tijd. Je experimenteert met de productie en export mogelijkheden. Een klant wil een nieuwe drug met specifieke effecten. Je maakt en test prototypes. De overheid start een campagne tegen een specifieke drug. Jij bedenkt een nieuwe

Gedurende de 5 jaar van de opleiding wereld burger met een sterke basis ken in moderne talen en een sterke tren Tijdens de opleiding zal je werken aan e deze werk je uit samen met je medes eerste jaar zal een introductie zijn aan d aspecten van de snel groeiende markt. M op verschillende excursies om te zien hoe in het echt uitziet en hoe het werkt. Je zal

The design required the academy to take a position in relation to the problem of drug trafficking. The design team actually engaged in the discussion with the academy management and visualized it in concrete terms. One could imagine that establishing such a program as an experiment could have quite a disruptive effect on the current policy on drug trafficking in the Brabant province. Or that such a draft proposal could create a different kind of involvement in the problem in an unexpected way. Much constructive discussion took place in the various design teams about the social role that they see for the designer, whom or what to confront, and how to determine the purpose of the design proposals.

Example of educational project 'Void'

Another project illustrating this type of applied design research is 'Void' at CMD Breda. In this project, students explore alternative and speculative forms of interaction in order to criticize existing conventions and systems.[30] During Void's last iteration in the fall of 2020, 120 sophomores researched a speculative future in which "man" is no longer dominant. By asking what-if questions, they explored the speculation of a post-anthropocentric world and its consequences for the relationships between human and non-human entities and systems. In addition, students mapped out the dominant systems that structure our contemporary human-centric worldview and tried to develop alternatives. The provocatypes **31** **32** (prototypes with the purpose of stimulating ethical imagination and discussion) that they then created offered a fragmentary experience of what a post-anthropocentric world could feel like as well as stimulating critical reflection on the present.

An example of this is the project 'Slave to the Rhythm' (Figure 3). In this study, Esmay Klein researched a future in which 'smart devices', under the guise of convenience, are taking more and more agency away from people. She asks the critical question of what 'outsourcing' more and more activities to smart technology means for human autonomy. Her design of a set of dysfunctional hair combs allows users to experience that even the most mundane activities can suddenly take a lot of effort when people no longer make independent decisions and determine how things work.

In this and other cases within Void, speculative design is not a form of trend watching, trying to predict the next innovation. On the contrary, it is a tool to question the socio-cultural and ethical consequences of technological developments and to use design to show that the world can also be different. Moreover, by imagining desirable as well as undesirable futures, this type of design research can help to determine actions that need to be taken in the present to create a desirable future. In this sense, applied design research enables students to better understand how the present is creating the future ('futuring') thus actively trying to 'critically challenge the natural order of things'.

30. James Auger, "Speculative Design: Crafting the Speculation," *Digital Creativity* 24, no. 1 (March 2013): 11–35. https://doi.org/10.1080/146 26268.2013.767276.

31. Cennydd Bowles, *Future Ethics* (London: NowNext, 2018).

32. Carl Disalvo, *Adversarial Design* (Cambridge: The MIT Press, 2012).

Figure 3
In the 'Slave to the Rhythm' project, Esmay Klein designed a series of dys-functional combs. By using them, users experience that even the most mundane activities can take a lot of effort when they can no longer control how things work.

33. Hartmut Rosa, *Resonanz: Eine Soziologie der Weltbeziehung* (Frankfurt am Main: Suhrkamp, 2016).

34. Hartmut Rosa, *Unverfügbarkeit (Unruhe bewahren)* (Salzburg: Residenz. 2018).

35. This is the essence of Caradt's mission statement as well. Compare: https://caradt.nl/about.

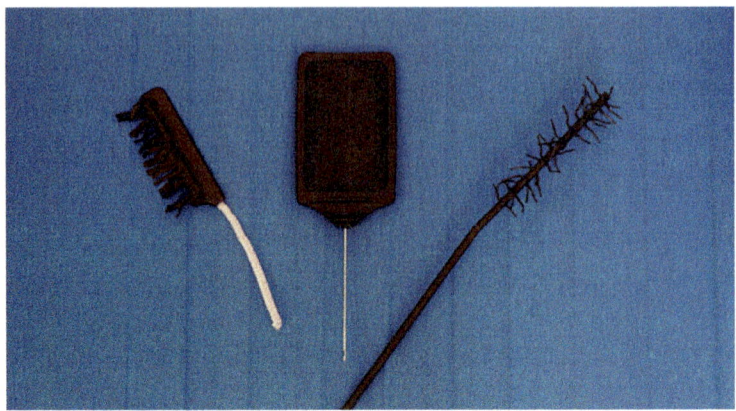

Further research

We will try to expand and strengthen this type of design project in the coming years by, among other things, working with other (creative) courses within and outside Avans University of Applied Sciences. Within design programs and in design practice, we want to focus on ways to allow for further resonance [33] [34] and provide room for more critical orientations on design research. If designers want to build a sustainable future, they also need infrastructures through which the current practice can be thoroughly questioned and the ideological chaff can be separated from the progressive wheat.[35]

In this context, it seems relevant to us to examine how forms of social criticism, more autonomous practices and other forms of collective organization and vision formation can strengthen each other in the design research, in order to increase their effect or impact in social and societal terms. In short, we see a great urgency to make this type of applied design research a regular part of the student orientation in the design field, and also make it part of how designers can manifest themselves in the creative domain in the future.

Eke Rebergen, Wander Eikelboom & Sebastian Olma

Avans University of Applied Sciences

Eke Rebergen and Wander Eikelboom both work as lecturers of CMD courses at Avans University of Applied Sciences, and as researchers within the Cultural and Creative Industries research group of the Center of Applied Research for Art, Design and Technology (CARADT). They are involved in curriculum developments within the courses based on their experience in the design field (Wander, among other things, as part of the Polymorf design collective). Sebastian Olma is a professor in this research group and is also a board member of the national platform Kunst ~ Onderzoek. He received his PhD from the Center for Cultural Studies, Goldsmiths College, University of London in 2007 and subsequently conducted research at the Amsterdam University of Applied Sciences' Institute of Network Cultures.

Discomfort as a starting point

How design research can contribute to design practice

Marieke Zielhuis

As researchers, how can we ensure our *design research* projects contribute even better to something designers can use in practice? That is the question I have been focusing on. My interest in this topic has grown in the years that I was active as a project manager within the research group Co-Design at the Utrecht University of Applied Sciences. In many of our projects – large consortia with academic and practical partners – we tried to produce valuable insights to benefit the *design professional*. Think of a service designer at a small agency, a healthcare product designer, or an interaction designer at a major digital agency.

To this end, we developed output such as practical tools and card sets (see an example in Figure 1). That was quite a challenge at times. For example, what is a suitable and practical form that designers can use in practice? Which content do they need or want? Also, it can be hard for researchers to allocate time and money to bring research results into practical use. Much attention is focused on developing solutions for people who are directly involved in the problem context, such as healthcare workers or the elderly. But the delivery

Figure 1
Example of an attempt to provide useful results for design professionals. This card set was developed in collaboration with professional designers as part of the SIA-funded research project Touchpoint. It helps designers gain a clearer view of their target group's behavior in designing for behavioral change.
Photo: Marieke Zielhuis. huis).

of concrete output that other designers could also use in such a context is considered less important. That could be a missed opportunity.

In addition, I find it uncomfortable that the designers we collaborate with sometimes have to invest more in a project than they can get out of it for themselves. In many projects, designers apply their practical expertise in user research or prototype development, or they act as a client or supervisor of design students. The insights they gain during their engagement are illustrative for what could be relevant for a broader group of professionals. For example, they develop a better understanding of a particular target group or familiarity with a theoretical model.

However, due to the often-limited role of the designers, these insights are limited to a part of the project; the designers are not included throughout the project. And those insights do not always reach a broader group. And finally, the participating design professionals often do not see themselves and the broader design practice as a target group in such a project; they mainly focus on contributing to

116

a societal problem. In such projects, where the design professionals mostly play a facilitating role and where design practice is hardly served as a target group, they are still asked to fund (part of) the effort they put in.

This motivated me to focus on this topic and investigate the following research question: *How can researchers reinforce the methodology of their applied design research projects so that more knowledge is developed that is useful for design professionals?*

1. Daan Andriessen, *Praktisch Relevant én Methodisch Grondig? Dimensies van Onderzoek in het Hbo,* Public Lecture at the Utrecht University of Applied Sciences (10 April 2014).

Studying research impact

The research group Research Competence at the Utrecht University of Applied Sciences provided the opportunity for me to carry out this research as a PhD study, in collaboration with the department of Industrial Design at TU Delft. The research group includes, apart from myself, researchers who are also interested in 'research into research', in professionalizing research and in realizing relevant contributions to practice. The research group focuses on the methodology of practice-based research in all disciplines in higher professional education, ranging from technology to education and from healthcare to the arts. We use the term *practice-based* research to indicate the type of scientific research that arises from concrete issues in practice, is carried out in close cooperation with practice, and has the explicit goal of generating relevant knowledge that can be used to support practice.[1] One of the central questions within this research group is: How can we ensure that the results of practice-based research have an impact on professional practice? In my own research, I narrow this question towards the practice of professional designers.

Applied design research

To answer my research question, I study various research projects (within universities of applied sciences and universities) that I would label *applied design research*. They are all practice-based (*applied*): arising from concrete practical issues, carried out in close collaboration with the practical environment and with the explicit objective of

2. John Zimmerman, Erik Stolterman and Jodi Forlizzi, "An Analysis and Critique of Research Through Design: Towards a Formalization of a Research Approach," in *Proceedings of the 8th ACM Conference on Designing Interactive Systems – DIS'10* (2010): 310–319.

3. Pieter Jan Stappers and Elisa Giaccardi, "Research Through Design," in *The Encyclopedia of Human-Computer Interaction, 2nd edition, eds.* Mads Soegaard and Rikke Friis-Dam (Aarhus, Denmark: 2017): 1–94.

4. Kees Dorst, "Design Research: A Revolution-Waiting-to-Happen," *Design Studies* 29, no. 1 (2008): 4–11.

5. Don Norman, *Living With Complexity* (Cambridge, MA: MIT Press, 2010).

6. Johanneke Minnema, Lisa Rosing, Marjolein van Vucht, eds., *Veerkracht – Kennis- en Innovatieagenda voor de Creatieve Industrie 2020–2023* (Eindhoven: CLICKNL, 2020).

providing relevant knowledge for practice. My definition of design research here needs some additional explanation because the terminology in this field can be confusing. The term *design research* is commonly used to 1) indicate how research is done (also in practice) for the purposes of a design project, to 2) indicate research that contributes to the continued development of the design discipline, and to 3) indicate research that has a design approach.
I study research projects that aim to combine the last two characteristics.

Within the third category, different traditions can be distinguished. I am most familiar with the tradition of design research in which designing and creating prototypes is seen as an indispensable part of the knowledge development, internationally mostly referred to as *research through design*. [2] [3] *Research through design* has its origins in the disciplines of arts, design and architecture, and has shown strong developments, in particular, the discipline of Human-Computer Interaction. It is a relatively young discipline, grown from the need of the design community to establish its own research culture with a more academic foundation.

It has since gained a fairly central place in the larger field of *design research*. In the discourse on *research through design*, different trends can be distinguished, with differences in cultural context and academic foundation: focusing on technical universities in the Netherlands, on art and design in the United Kingdom and Scandinavia, and on HCI/IxD in the United States. In the Netherlands, the NWO call 'Research Through Design' highlighted such designing forms of research in 2014.

I would also consider my own research approach in this PhD project as a form of *applied design research*. The practical part (*applied*) shows in my aim to deliver practical tools: tools that enable researchers to strengthen their projects in their practical contribution for design professionals. It also refers to the collaboration with my own practice audience: researchers as well as professional designers.

My research is *design research* in its contribution to the academic design field and ultimately to design profession-als (i.e., the second meaning of the term), but also in the way I incorporate the design of interventions and tools in a *research through design* approach to develop knowledge (the third meaning). I am not only interested in the result-ing interventions and tools, but also in the insights that result from the design processes that I want to engage in with researchers and professional designers. In addition to this designing part, a large amount of my research consists of case studies, in which I study *design research* projects (through interviews and document analysis), but in which I do not design or intervene myself.

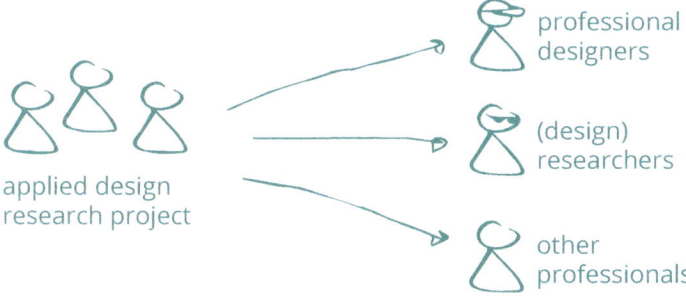

Design practice as a target group: Why is it challenging?

Figure 2
The various audiences in an applied design research project.

What makes it so difficult for researchers to contribute to designers in practice? *Applied design researchers* usually want to contribute to practice, not just to science. This is even an explicit task for researchers at universities of applied sciences. The need to develop knowledge for the design practice is there: the changing roles that designers need to play require new knowledge.[4] [5] Various relevant research areas are described in the knowledge and innovation agenda [6] by CLICKNL, the organization that represents the interests of the creative industry in the Netherlands.

What makes it difficult is that the current subsidy landscape in the Netherlands does not offer much opportunity to develop knowledge directly intended for design practice. Subsidized projects are tasked to focus primarily on social issues, such as healthcare or sustainability. Contributing

119

7. Yvonne Rogers, "New Theoretical Approaches for HCI," *Annual Review of Information, Science and Technology* 38 (2004): 87–143.

8. Marieke Zielhuis, Froukje Sleeswijk Visser, Daan Andriessen, Pieter Jan Stappers, "What Makes Design Research More Useful for Design Professionals? An Exploration of the Research-Practice Gap," *Journal of Design Research* (in press).

to design practice is often not supported and is not something that researchers are assessed on. This means that the knowledge base of the creative industry is developed primarily in the slipstream of projects with a different primary objective. This means that we should take full advantage of any opportunities, *especially* in those projects where design professionals are not the most important target group.

These opportunities will have to be sought in projects with multiple target groups with a wide range of interests. A single project can have an audience of researchers, healthcare professionals, government employees, entrepreneurs, design professionals, and people with dementia. Figure 2 shows the main different audience groups of applied design research projects. I have observed that combining these interests can be a challenge: the target group of design practice is easily overlooked in relation to the primary target group, such as healthcare workers or the elderly. If solutions are aimed at practice, they are primarily aimed at these target groups. And when we aim at a designer audience, we sometimes respond – consciously or unconsciously – more to the interests and needs of (design) researchers or (design) students than to the practice interest of professional designers.

Researchers also regularly find that the methods they developed are not used in practice as they had intended.[7] Design professionals have different interests and preferences than academic researchers, even if the latter have a design background.[8] The interviews I conducted among design professionals indicated that they indeed do not use tools – such as the above-mentioned card set – as intended by the research team. However, these tools prove very useful for them to demonstrate and illustrate the application of an underlying theory.

Collaboration with real-world designers

One way to bring in a practice perspective is to collaborate with partners from practice. The creative industry community in the Dutch context is already relatively closely

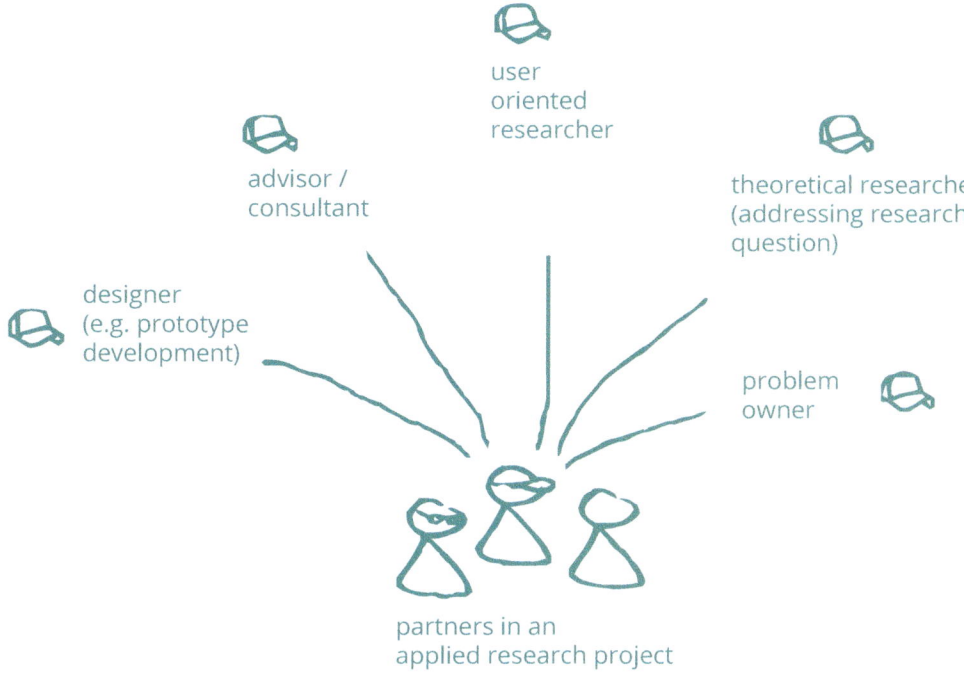

user
oriented
researcher

advisor /
consultant

theoretical researcher
(addressing research
question)

designer
(e.g. prototype
development)

problem
owner

partners in an
applied research project

connected to the design research community at universities and universities of applied sciences. I see that design
professionals are involved in research projects in various
roles: 1) as a designer (e.g. of prototypes), 2) as an advisor,
for example as a senior design coach for student teams
or in a sounding board, 3) as a hands-on researcher, e.g.
conducting user research, 4) as a theorizing researcher, and
sometimes 5) as a problem owner (see Figure 3).

The project that produced the card set in Figure 1 involved
professional designers in a combination of roles 3, 4, and
5. This was one of the few projects in which we could
prioritize design professionals as a target group. In many
cases, however, such research-practice collaboration is
primarily geared toward social goals, not toward the needs
of design practice. And that doesn't sit right. This collaboration is challenging when you, as a designer, have a role as a
co-troubleshooter in a project but are funded as if you are
the problem owner (and have to co-fund the project).

Figure 3
Various roles for design
professionals as partners in
an applied design research
project.

121

9. Jonas Löwgren, "Annotated Portfolios and Other Forms of Intermediate-Level Knowledge," *Interactions*, (February 2013): 30–34.

10. Bill Gaver and John Bowers, "Annotated Portfolios," *Interactions 19, no.* 4 (2012): 40–49.

Promising developments

The uncomfortable feeling that I described at the beginning of this article is linked to the paradoxical position that professional designers seem to take in *design research*: on the one hand, they are seen as co-solvers of a societal challenge; on the other hand, there is little opportunity to develop knowledge that enables them to function optimally as problem solvers.

In light of these challenges, I see some promising developments. One of these is the growing focus on suitable formats to bring the insights and experiences from design research to design practice. A wide range of intermediate forms has been identified between abstract knowledge and concrete design solutions.[9] For example, *annotated portfolios* [10]: collections of design solutions in which the annotations provide a better picture of, for instance, design decisions. In addition, funding providers are gradually offering more opportunities to consider design practice as a target group, such as SIA's GO-CI program.

Funding providers also offer more opportunities to involve designers from the field in research projects. A significant challenge is to develop ways to co-create with design practice and to share knowledge and best practices about these ways. Especially in large, long-term projects with many research and practical partners, the question must be how designers can find a role that matches the dynamics of their practice and in which the give-and-take is well-balanced.

Finally, it seems a challenge in itself to reflect on such matters, to develop best practices and to share them with others. Especially in a field such as *applied design research* with a multitude of approaches and concepts. I expect that a further development of the shared definitions framework (or understanding of the differences) and an overview of the varied landscape of approaches in the *applied design research* field will support such reflection and contribute to a more powerful contribution to the practice of professional designers.

Marieke Zielhuis

Untrecht University of Applied Sciences

Since 2018, Marieke Zielhuis has been a researcher at the research group Research Competence at Utrecht University of Applied Sciences. She focuses on methodical challenges within applied design research on developing knowledge for design practice. In 2019, she started a PhD research project in collaboration with the Industrial Design department at TU Delft. Trained as an industrial designer (TU Delft), Marieke has worked at the Utrecht University of Applied Sciences since 2002. She has been the project manager for several research projects within the Co-Design research group. She has also played a role in two Centers of Expertise: UCREATE (Creative Industries) and Smart Sustainable Cities.

123

PART 3: DESIGN AND RESEARCH WITH OTHERS

"In a fast and profoundly changing world, everybody designs. "Everybody" means not only individual people, groups, communities, companies, and associations, but also institutions, cities, and entire regions"

~ **Ezio Manzini**

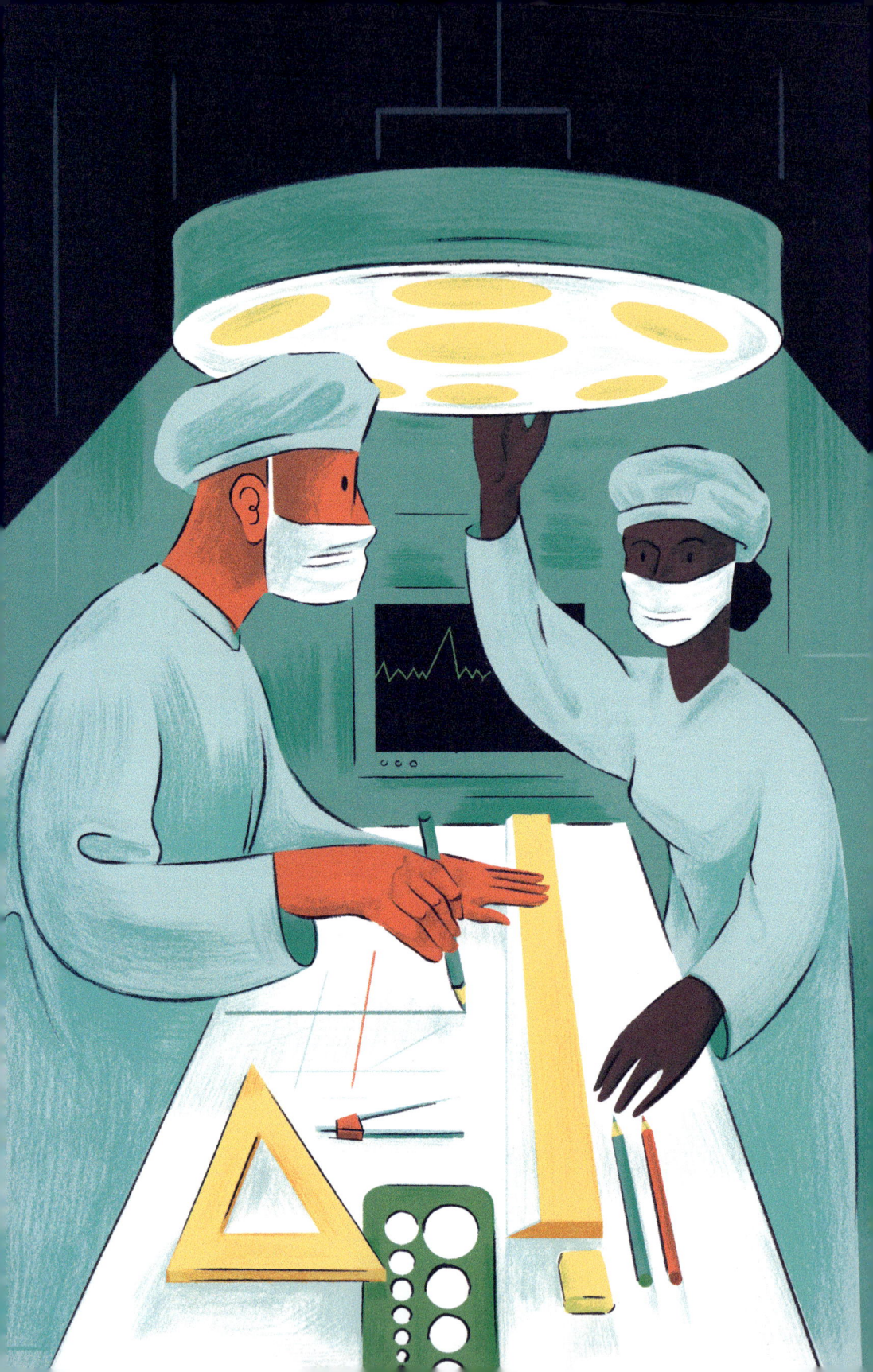

Systemic co-design
The designer as a facilitator of change
Remko van der Lugt

There I was, heart pounding, at the Emergency Department of a large hospital. With an interdisciplinary group of product, service, experience, and interior designers, we had come to the end of an intensive week, during which we, together with the staff of the Emergency Department, identified opportunities for innovations. In the coffee room, we had created a map with insights and discussed the results with the nurses and team leaders. It felt like a privilege to be admitted to this strange working environment, where you usually only get a patient's view of how things work.

As a team, we noticed that we had become so alert to the subject during that week that every time we heard an ambulance siren, we looked up, ready to jump on it, hungry for new insights. It actually felt a bit wrong, as if we were ambulance-chasing lawyers. Curiosity is essential to designers, but it is equally important to be empathic when approaching people and situations. We also realized time and again that the ambulance was on its way to real people, in real distress.

During the design week, we worked with and alongside the nurses while they were doing their job. We examined, observed, talked to the staff and patients about their experiences, generated ideas, and developed them in draft prototypes, to try them out and evaluate them with the team straight away. In the beginning, the nurses were pretty skeptical. They were used to having researchers or consultants hanging around in the corridors. In most cases, they said that these researchers had a certain distance or bias about how the work should be done, without listening to the knowledge and experience of the nurses. Fortunately, the staff quickly warmed to our group. They appreciated our open and modest attitude as we entered their working environment, and they felt heard and inspired.

Figure 1
Discussing the 'innovation map' with the nurses.

In the end, the week resulted in several innovation directions. From quick wins such as a way to secure breathing equipment cables so that they couldn't get disconnected by accident, to new work processes so that cancer patients don't routinely have to go through emergency care, because the occurrence of their side effects is often predictable. From simplifying the flow of information from the ambulance through the reception desk to the trauma room to improved experiences for patients, such as alleviating the long waiting period or creating an appropriate environment in a room to say their farewells when their next of kin is not going to survive.

I was terribly impressed by the nurses' actions, their perseverance, and their problem-solving capacities! They were great at solving immediate problems they encountered and finding workarounds in the big system. However, I was surprised by their somewhat cynical attitude toward innovation. They had often tried to initiate innovations themselves, but they were repeatedly hindered in their efforts by the organization's system. On the other hand, new products were added to the department frequently, but they did not really suit the nurses' real needs and working methods. The need to give these professionals an equal voice in developing the products, the processes, and the work environment became clear to me once again!

128

As a professor of Co-Design at the Utrecht University of Applied Sciences, I have been immersed in ever-diverse working contexts with an enthusiastic and diverse group of designers-researchers since 2013. Part of my job is introducing people in these contexts to design collaborations. It is that broad range that makes the work so exciting. One moment you are in a hospital; the next moment, you are working at an IT company, a paramedical agency, a school, at a construction site, or working for the government. As a research group, we participate in various research projects that always focus on complex social innovation. Our research involves developing design capabilities in individuals and teams: what do they need in skills and tools to participate in a joint design process fully? And what does that require of the designer as a facilitator in this process?

1. John Heider, *The Tao of Leadership: Lao Tzu's Tao Te Ching Adapted for the New Age* (Atlanta, Georgia: Humanics New Age, 1986).

Figure 2 & 3
Exploring a new information path from the ambulance to the trauma room with the nursing staff.

We are increasingly aware that complex issues require a systemic attitude, are sensitive to – and find a foothold in – often hidden dynamics in the system. Within this structure, we connect the designer's tools, methods, and attitude to complex thinking and systemic working.

I am often inspired by the metaphor of the designer as a midwife.[1] It is about creating a safe space for people to gain insights and to conceptualize themselves. Hard intervention is sometimes needed, but always based on the conviction that the process is owned by those who will live with the new design.

Facilitating co-design

Some years ago, I facilitated a creative design session for a large telecom technology company. The participants included business developers, marketers, technology experts, and designers. The company had flown in a hotshot stylist designer from the US to inspire and empower the participants during the session. This designer was a great sketch artist and was able to convey his vision of the new design. The problem was that this somewhat intimidated the participants. In fact, they did not dare to put any more ideas on paper or put forward ideas.

In facilitating co-design processes, that is the actual challenge. How do you enhance participants in their design ability so that they can contribute fully? And how can you, as a designer, take on a role of service to that collaborative process while also engaging your own creativity? Facilitating co-design processes requires a rather delicate balance of guiding participants through the process and enabling them to create their own path through training and/or providing materials that will allow people to think and act in a 'design way'. We often tailor these materials specifically for the context and characteristics of the task and the participants.

In general terms, these materials can be divided into two groups. On the one hand, the materials that enable participants to act in a design manner, somewhat ending the creative dominance of the designer. These are, for example, graphic elements for collages or kits to develop models and prototypes. Here, we rely primarily on the accumulated experience with generative techniques and context mapping.[2] On the other hand, we design materials that make the already accumulated content knowledge accessibly and actively available in the co-design process. These include knowledge cards, personas, and posters such as timelines and infographics.

Systemic co-design

For me, systemic thinking is not necessarily about large-scale thinking. Even small-scale societal problems often hide a world of complexity and system dynamics. Systemic co-design focuses on these hidden dynamics between (groups or networks of) people, with the conviction that if you can see the system dynamics, you can also see where you need to focus your intervention as a designer.

2. Elizabeth Sanders and Pieter Jan Stappers, *Convial Toolbox. Generative Research for the Front End of Design* (Amsterdam: BIS Publishers, 2012).

The recent project 'What Moves You?!' (Wat Beweegt Jou?!) in the field of pediatric physiotherapy is a good example. Children with disabilities tend to exercise too little, especially in their daily environment. There are several reasons for this. Parents or coaches are afraid that something will happen to the child, or neighbors do not know how to behave toward the child. In a co-design process with pediatric physiother-apists, community sports coaches, parents and children (Figure 4), we developed a toolkit with easy-to-use interven-tions. For example, think of a small plexiglass window that forces a meddling parent to take a more distant observing role during the treatment, giving the child room to talk and try things out. Or think of a picture frame that lets the child show people, for example their PE teacher, what they can do and what they need in support (Figure 5).

Figure 4
What Moves You?! Toolkit with different interven-tions used in paediatric physiotherapy.

Co-research through co-design

The co-design facilitation attitude also translates into how we conduct research within the research group. Our approach is participatory, involving participants in design projects as co-researchers. We use generative techniques to enable the project team to gather data and generate insights throughout the process. For example, reflective journals help keep an eye on the research question and record everything encountered. We also often use interview posters designed for the situation. These enable co-researchers to conduct in-depth conversations with people in their own environment (colleagues, neighbors, family members, etc.). We then interpret the collected data with the co-researchers, actively using the entire space (the walls, the floor) to assist in this process.

Figure 5
Model of a picture frame that allows a child to show, for example to the gym teacher, what he or she knows and what support he or she needs to achieve it.

This type of research often motivates people; it makes the participants feel involved in a collaborative learning process. However, there must be sufficient trust, and if there are interpersonal or organizational problems, these are addressed first. Applying co-design when a reorganization is underway, and people are afraid to lose their jobs is destined to fail.

132

Reflecting on applied design research

For me, the term 'applied' in applied design research refers to researching while we are designing in, for, and with the chaos of reality. The aim is to deal with the pragmatics, dynamics, and pace of design in the real world that goes with it. The challenge is to add a layer of systematic knowledge development to this practice without affecting the design process too much. With applied research-through-design, we see the design process as an explicit knowledge-generating activity, not separate from the research. These can be insights from collecting and organizing smaller and larger considerations when designing an intervention. For example, what was the reason for that change from a clean design to a warm and friendly design? Why do we think that button should be placed there? What other solutions did we have in mind? And what was the reason we discarded those? The 'applied' aspect presents several challenges that we can address together with the NADR network and that can also benefit the broader design and research community.

Figure 6
Collective sense-making: Interpreting the interview results by organising them in a landscape on the floor.

Reflection-IN-action as a knowledge-generating engine

As described above, in research-through-design, the design team members also participate in the research as co-researchers, who, during the design process, are sensitive to gathering information about the research question. For example, a speech therapy project that covers the following questions: How can you design for the social dynamics

3. Donald Schön, *The Reflective Practitioner: How Professionals Think in Action* (New York: Basic Books, 1984).

4. Joel M. Hektner, Jennifer A. Schmidt, Mihaly Csikszentmihalyi eds., *Experience Sampling Method: Measuring the Quality of Everyday Life* (Thousand Oaks, CA: Sage Publications, 2006).

5. George Roth and Art Kleiner, *Field Manual for a Learning Historian* (Boston: MIT, 1996).

between the parents, the child, and the speech therapist? When we apply Donald Schön's distinction between reflecting IN action and reflecting ON action,[3] it is not difficult, as a team, to look back on the design activities and to gather relevant insights.

However, this usually leads to rather broad views on reflections. It lacks the refinement needed to gain insight into the micro-steps and decisions taken in the process, while it is precisely there that many valuable insights can be found. That requires reflection IN the design process by the participants. However, this is not as easy as it seems. It is challenging to maintain both the creative flow of the process and at the same time record the reflections. One possible developmental direction is to make reflection-in-action extremely easy and accessible, for example, by applying principles from experience sampling .[4] In specific reflection-on-action moments, designers can, for instance, use the many micro-insights to create a learning history [5] or project journey map. This approach should be further strengthened as a research methodology. Together we can work on a repertoire of smart research tools that make applied design research even more accessible through reflection-in-action.

Balancing the rigor of research and the dynamics of practice

The practical nature of our research brings with it grand ambitions and limited budgets, and opportunities to spend time solely on data collection and interpretation. Design in and with practice is usually done in fast iterations, making it difficult to do enough solid research before, during, and after the design process. This usually makes it challenging to publish insights into domain-content research journals (in our case often paramedical in nature) that are used to predict certain research methodology predictability. How can we find appropriate ways to serve both the requested rigor of domain scientific research and the speed needed for practical relevance in the practical project?

134

Applied design research as a methodology for systemic change

The design field is increasingly moving into the field of systemic transition or transformation processes, intending to bring about change in complex socio-technical systems. We are becoming increasingly aware that applying and participating in co-design and co-research itself is already a way of contributing to such changes, not just the outcome of that process (the resulting designed services, products or interventions on the one hand, and the resulting knowledge on the other). NADR member Perica Savanović, lecturer of Built Environment at Avans, made me aware of this. Performing co-design and co-research together with the people in the relevant societal domain can initiate much movement. It is also essential to further develop this function of applied design research as a key enabling methodology and collect a repertoire of cases to draw from.

Remko van der Lugt
Utrecht University of Applied Sciences

Dr.ir. Remko van der Lugt investigates how the tools, methods, skills, and attitudes of designers can accelerate complex societal innovation projects. He focuses on how to engage people fully as experts of their own experiences. He links the range of thoughts of participatory design to systemic thinking and develops working methods that enable designers to operate as facilitators of change. Examples include Gigamapping, Design Probing, Stakeholder constellations, and Socionas. Remko obtained his PhD at TU Delft. Since 2007, he has been a Professor of Co-Design at the Utrecht University of Applied Sciences, where he also serves as co-director of the Leren en Innoveren (Learning and Innovating) Research Center.

Inclusive designs in healthcare

Warm Technology for people with dementia

Rens Brankaert

Within my research, the perspective and experiences of the target group we design for are the most critical elements. We follow a human-centered design approach, where the target group is part of the design process.[1] When designing new technology in healthcare, the medical or technological perspective is often dominant, while the client or patient perspective is often overlooked.[2] The importance of inclusive and human-oriented work in healthcare is therefore becoming increasingly important, building on the concept of person-centered care.[3]

In this person-centered design approach, we actively involve target groups in the design process through participatory methods such as co-design.[4] As such, they support making decisions and co-create the value that new technologies and services can deliver.[5] By collaborating with people, we ensure that new technologies and services contribute to improving the well-being and quality of life.

Involving the target group in design processes is more challenging in vulnerable groups, such as people with dementia. However, doing so is crucial so that appropriate

1. Tom Kelley, *The Art of Innovation: Lessons in Creativity From IDEO, America's Leading Design Firm,* (New York: Crown Publishing Group, 2007).

2. Amanda Lazar, Caroline Edasis, Anne Marie Piper, "A Critical Lens on Dementia and Design in HCI," in *Proceedings of the 2017 CHI Conference on Human Factors in Computing Systems* (Denver: ACM Press, 2017), https://doi.org/10.1145/3025453.3025522.

3. Rens Brankaert, Gail Kenning, Daniel Welsh, Sarah Foley, James Hodge, David Unbehaun, "Intersections in HCI, Design and Dementia: Inclusivity in Participatory Approaches," in *DIS 2019 Companion – Companion Publication of the 2019 ACM Designing Interactive Systems Conference* (San Diego, CA, June 2019), https://doi.org/10.1145/3301019.3319997.

4. Rens Brankaert and Elke Den Ouden, "The Design-Driven Living Lab: A New Approach to Exploring Solutions to Complex Societal Challenges," *Technology Innovation Management Review* 7, no.1 (2017): 44–51.

5. Elizabeth Sanders and Pieter Jan Stappers, "Co-creation and the New Landscapes of Design," *CoDesign* 4 no. 1 (2008): 5–18, https://doi.org/10.1080/1571088-0701875068.

6. Niels Hendriks, Karin Slegers, and Pieter Duysburgh, "Codesign With People Living With Cognitive or Sensory Impairments: A Case for Method Stories and Uniqueness," *CoDesign* 11 no. 1 (2015): 70–82.

7. James Hodge, Kyle Montague, Sandra Hastings, Kellie Morrissey, "Exploring Media Capture of Meaningful Experiences to Support Families Living with Dementia," in *Proceedings of the 2019 CHI Conference on Human Factors in Computing Systems* (May 2019): 1–14, https://doi.org/10.1145/3290605.3300653.

8. Wijnand IJsselsteijn, Ans Tummers-Heemels, Rens Brankaert, "Warm Technology: A Novel Perspective on Design for and With People Living With Dementia," Rens Brankaert and G. Kenning eds., *HCI and Design in the Context of Dementia* (Cham: Springer International Publishing, 2020): 33–47, https://doi.org/10.1007/978-3-030-32835-1_3.

solutions can be developed for that group too. We can do that by adapting our methods and making them suitable and accessible for them.[6]

Dementia is a general term for all kinds of conditions that progressively affect cognitive functioning. As a result, people become dependent on support from others and healthcare services. Moreover, people with dementia are often discussed, but not consulted. And sometimes, people with dementia are still considered to be *the disorder*, rather than the *individuals* they are, with their own wishes and needs.[7]

The statement 'Not about us, but with us', which was introduced in the 1980s by accessibility movements, still speaks for itself, especially when it comes to design in healthcare. Unfortunately, it is still not implemented on a large scale. In this chapter, I will illustrate how this can be done with complex target groups through a number of examples.

In Dr. Manon Peeters' Wearables Project, this approach is applied to wearable sensors. In this project, Manon investigates with colleagues and students how wearable sensors can be used meaningfully and appropriately to support people in complex healthcare (Figure 1), particularly when people with a need for care can no longer express themselves well. For example, this wearable technology can measure stress and stress-related problems almost in real time, allowing healthcare professionals to act proactively and appropriately, for example, to prevent escalation. In addition, such a wearable sensor could also be used to provide the target audience with easy access to personal and meaningful media, such as the dog picture shown in Figure 1.

Sometimes portable sensors are considered restrictive because they track the person receiving care and record all their movements. That is why it is essential to explore this technology in its context, together with the target group, and look carefully at how these sensors can play a meaningful role in healthcare and contribute to well-being and quality of life.

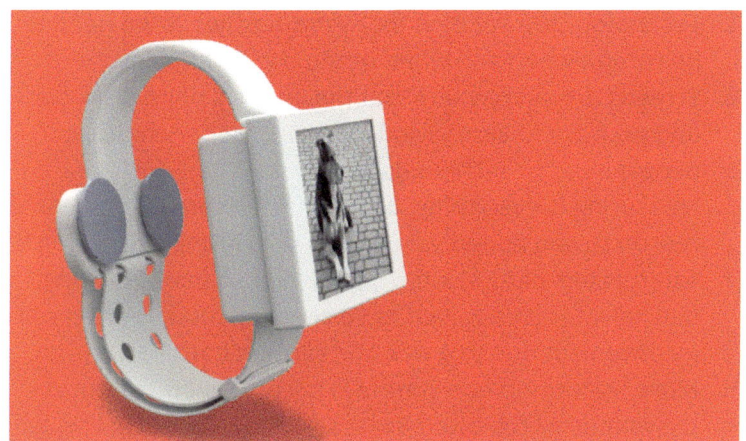

Figure 1
A wearable for complex care (photo Marko Hofman, student Fontys IPO).

This research by Dr. Manon Peeters takes place in practice, with healthcare partners and the people who work there, to address the challenges those partners have. If the technology is found to be appropriate, it can be applied in the day-to-day care of people. The next steps in the project are to better understand the context and to specifically explore when and how stress occurs. We can then develop appropriate services for this and improve complex care for groups that have challenges with expressing themselves.

Warm Technology

Over several studies and projects, we have developed the concept of Warm Technology as a vision that supports the design of inclusive technology in healthcare. Warm Technology focuses on what a person still can do, addressing both social and emotional needs. The technology is personally empowering, non-stigmatizing, easy to use, and fits in their social context.[8] Warm Technology intends to create technology that is more accepted, better addresses needs, and contributes to improve quality of life.

The concept of Warm Technology moves away from the 'temptations' that technology developers often face. In recent work we can see that developers seem to think the following:

139

9. Maarten Houben, Rens Brankaert, Saskia Bakker, Gail Kenning, Inge Bongers, Berry Eggen, "The Role of Everyday Sounds in Advanced Dementia Care," in *Proceedings of the 2020 CHI Conference on Human Factors in Computing Systems* (April 2020): 1–14, https://doi.org/10.1145/3313831.3376577.

10. Roger L. Martin, *The Design of Business: Why Design Thinking Is the Next Competitive Advantage* (Boston, MA: Harvard Business Press, 2009).

1. Technology is a solution for everything, while we need to be careful when and when not to use technology.
2. Screens are everywhere, but the world is not a glass plate and we have to accommodate the sensory richness of people into our technology.
3. Measuring is everything, but we need to provide a clear reason to and added value when measuring something.
4. 'Interpreted' natural interaction is the solution. Existing natural interaction, such as voice interaction, is often still inconvenient due to a lack of clear feedback.
5. More features in one system is better, we add features to systems all too often because we can. In doing this, we often unnecessarily complicate the technology.

The perspective of Warm Technology contributes to our mission to support people in care through technology. It is reflected in our design processes as well. This is very much apparent in the VITA project.

The VITA musical pillow was designed in collaboration with De Bende, Interactive Matter, and the PIT team as part of the healthcare organization Pleyade. In this design project, our goal was to make music and sound accessible again to people with dementia in long-term care. People with dementia often need 24/7 care support and experience serious cognitive and locomotor challenges. That is why traditional ways to listen to music, such as a CD player or a computer with Spotify, are not suitable for this audience.

VITA is the resulting design of an iterative and participatory design process that involved healthcare professionals, informal carers, and people with dementia. Figures 2 and 3 show the resulting design. People can play music and sound by placing their hand on one of the six fabric sensors (Figure 2). VITA is shaped like a pillow and inviting to touch, not intimidating, and easy to place on a person's lap, table or bed. It has a modern look, with high contrast, so the pillow stands out in the surroundings.

A simple interface on the back of VITA functions as the control panel, where caregivers can switch the power on/off, turn the volume up or down, and select a personal profile and a theme. The more complex settings can be managed

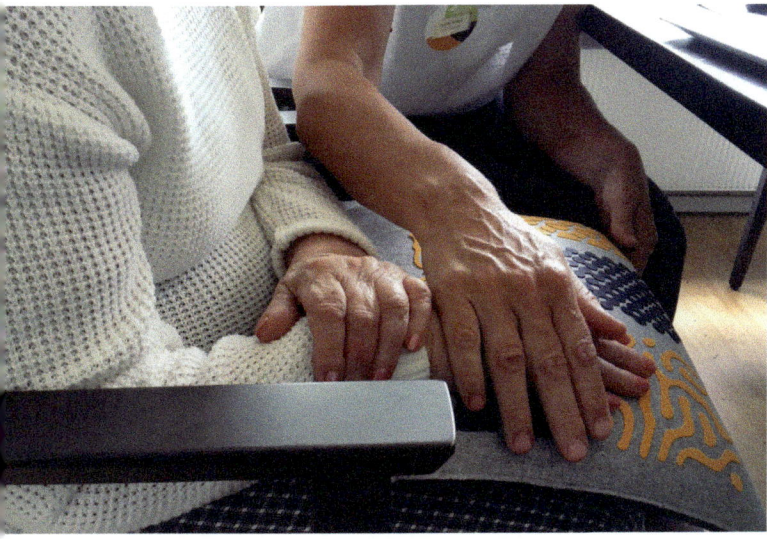

Figure 2
The VITA pillow in use as part of long-term dementia care.

through an app or online portal by both the healthcare staff and the patient's relatives. The profile function makes it possible to create personal music and sound sets so that all residents can have their own personalized sound or music experience with the pillow.

Our research showed that the VITA has added value in healthcare; it improves interpersonal contacts and thus creates a pleasant and valuable moment.[9] The VITA was used in several ways, sometimes by people with dementia independently, but more often with family members or care staff. The VITA was usually placed on the sofa or a cabinet after use (Figure 3), making the VITA available whenever people wanted to use it.

Applied design research

Design research is essential in the context of changing healthcare and the more inclusive involvement of people in healthcare. When it comes to significant social challenges such as dementia, the design discipline can bring new ways of thinking and new solutions. Designers have the skill to combine different perspectives into one concept.[10] In design research, the designers are both the inventors and the facilitators of innovation. A designer does this by identifying the existing needs and proposing concepts based on them. The designer then uses an iterative process to develop a

141

Figure 3
The VITA pillow as part of the environment in long-term dementia care.

prototype that is suitable for testing.[11] In addition, designers bring together different disciplinary insights using their sensitive and inclusive way of working.[12]

Both the research carried out by designers and the positioning of design as a research approach have led to Design Research becoming an academic discipline. According to Gaver,[13] one of the most important added values of this research field is 'the ability to challenge the status quo continuously and creatively'. Applied design research contributes in the same way to different sectors and professional practices. In the health sector, healthcare organizations and other stakeholders can benefit from this 'designerly' way of working and thinking.

One example that illustrates how this could work in practice is the Pleyade Innovation Team (PIT). The PIT initiative was set up at the healthcare organization Pleyade. The organization understood that innovation and change need to be done by professionals who have been trained to do so. Together with Pleyade, we have set up a team of healthcare professionals, design professionals, and design researchers. This team looks for latent needs within the organization and develops new concepts and ideas based on these needs.

Next, the team starts working in short design cycles to realize the concepts into experienceable prototypes that are then put into care practice. In this process, the organization learns from how designers work. It benefits their organization directly and has resulted in many new technological innovations, such as the VITA. The innovation team enables healthcare professionals to express their needs, invites them to think about innovation and different solutions to existing problems, and supports management to organize a future-proof healthcare.

Another example of a collaborative project between designers and health professionals is the collaboration between healthcare organization De Riethorst Driestromenland, design studio Luckt, and design researchers. A team of ten healthcare professionals was invited to design new Warm Technology within the healthcare organization. During the design process, we found a need that people with dementia were mostly too passive between planned activities, and sometimes even fell asleep. We went looking for a way to keep people with dementia active throughout the day.

Together with the team, we designed SAM (Figure 4): an interactive table friend that invites people to interact. During a participatory design process, the first SAM prototype was built and evaluated in long-term care. People can interact with the SAM by shaking, tapping, or caressing it. The SAM reacts with lights, sounds, and vibrations. The two SAM prototypes also interact with each other to create a social situation. For example, one gets jealous when the other is picked up and starts asking for attention.

In addition to the concrete product SAM, the results of the design process in terms of mindset and implementation were probably even more relevant. The team gradually became more open to new ideas and innovations, and more creative over the course of the process to tackle the found challenge with designers. In addition, the team embraced the SAM as a result of their own efforts. This contributed to the team spirit and the integration of the SAM into everyday care practice. This shows how design and designers contribute to the care domain as inventors and facilitators.

11. John Krogstie, "Bridging Research and Innovation by Applying Living Labs for Design Science Research," in *Lecture Notes in Business Information Processing* 124 (2012): 161–176, https://doi.org/10.1007/978-3-642-32270-9_10.

12. Julie Thompson Klein, "Prospects for Transdisciplinarity," *Futures* 36 no. 4 (2004): 515–526, https://doi.org/10.1016/j.futures.2003.10.007.

13. William W. Gaver, "What Should We Expect From Research Through Design?," in *Proceedings of the 2012 ACM annual conference on Human Factors in Computing Systems (May 2012): 937–946,*. https://doi.org/10.1145/2207676.2208538.

Figure 4
The SAM prototype to acti-
vate people with dementia.

Figure 4
The SAM prototype to acti-
vate people with dementia.

Advancing insight

Applied design research has grown considerably in recent years. This is demonstrated by a worldwide need for design-ers to contribute solutions for societal challenges such as dementia. In addition, designers' skills have also become interesting for non-designers. Example projects like the above show how organizations can benefit from including designers in their work practice. In my position at the Fontys University of Applied Sciences, these developments also take place: healthcare professionals develop design skills that are relevant to work with changing circumstances and new innovations in their own healthcare practice.

To professionalize applied design research, we need to explore the added value of this approach and the challenges we encounter in practice further. This could be done in at least the following three ways.

1. By teaching design skills to non-designers and training designers to include all relevant stakeholders in their processes as facilitators. In our research, we can identify, map, and build a repertoire of best practices in different domains for the processes for multidisciplinary and trans-disciplinary collaborations.

2. By stronger advocating, as design researchers, an inclu-sive way of working, as described in the introduction of this chapter. This subject needs to be discussed further. It will enable us to tackle prejudice and inequalities, which unfortunately still exist widely in our society. In the case

of dementia, for example, this has to do with age and stigmatization of the disease, but there is also room for much progress in other areas such as culture, diversity, and socio-economic status.

3. Finally, by looking for sustainability within applied design research. Sustainability in the sense of a long-term implementation and impact of design or process results, ensuring that target groups can benefit, now and in the future. If a change occurs only when designers are present (which means that innovation depends on the designer), it undermines the societal challenge and the necessary transformation. Only by designing, experimenting with, and creating a new way of working in professional practice can we change society and give people in healthcare a high quality of life and work.

Rens Brankaert

Fontys University of Applied Sciences

Dr.ir. Rens Brankaert is a professor at the Paramedic School of Fontys University of Applied Sciences and an assistant professor at the Eindhoven University of Technology. He researches how we can take a more person-centered and personal look at technology for people with dementia. He calls this 'Warm Technology', and explores this with the target group through co-design, Living Labs, and applied design research in healthcare practice. Rens Brankaert is a co-director of the TU expertise center Dementia & Technology, a Key Technology Partner Fellow at UTS Sydney, and the scientific director of the Dementia Lab Conference. In 2021, he won the Young Outstanding Researcher Award, awarded by the Alzheimer Nederland foundation, for his research on Warm Technology.

Societal impact design

Empathic and systemic co-design as a driver for change

Wina Smeenk

Societal challenges are becoming increasingly acute. As people, citizens, residents, and city users, we all have to deal with them. These are topics such as dementia, climate change, and COVID-19. It is difficult to truly understand and address these challenges, because there is not one owner; issues are interconnected, intertwined, and dynamic. Moreover, issues can fall outside the region, focus, tasks, and responsibilities of the stakeholders involved. Therefore, it is difficult to get a good overview of the situation, make joint decisions, and take steps together.

This demands both an individual and collective orientation to address these orphaned challenges that fall between the cracks. Nowadays, design is increasingly considered a possible approach to these kinds of challenges,[1][2] because design can handle uncertainty, is optimistic, and investigative in nature. Moreover, through its experimental and action-orientated nature, design can contribute to creating meaningful, alternative futures.

This thought has greatly broadened our design field over the last decade. From designing esthetic, functional products and services, designers are now increasingly committed to

1. Ezio Manzini, *Design, When Everybody Designs: An Introduction to Design for Social Innovation* (Cambridge, MA: MIT Press, 2015).

2. Daniela Sangiorgi, "Transformative Services and Transformation Design," *International Journal of Design* 5, no. 2 (2011).

developing meaningful experiences and work processes, and even making people aware of their influence on certain pressing situations through design (interventions).

Figure 1 (left)
The dementia simulator, exterior. Photo by Jacqueline Gielen.

Figure 2 (right)
The dementia simulator, interior. Photo by Jacqueline Gielen.

The Dementia Simulator

One example of such a complex challenge is dementia. People with dementia are often misunderstood. One of the main challenges is how family members, informal carers, and health-care professionals can better relate to someone with dementia to make both their (working) lives more enjoyable. I have worked on this issue in a multi-stakeholder coalition: a collaboration with healthcare institutions, the corporate world, knowledge institutions, people with dementia, and their partners (see Figures 1 and 2). Research using empathic discussions, simulations and role-play has provided new insights into strategy development and supported finding a shared ambition in the promising idea of a dementia simulator. Through a visit in the simulator, healthy people (healthcare professionals and informal carers alike) can experience what it is like to live with dementia. By experiencing this, they are emotionally 'touched', increasing their understanding. This then motivates them to look at their own behavior and adjust it where necessary. Ultimately, it improves the home situation: the simulator and subsequent training provide behavioral changes for individual informal carers and/or healthcare professionals, allowing people with dementia to continue to live at home longer and enabling professionals to work more comfortably and effectively.[3]

My brand-new research line Societal Impact Design at Inholland's Creative Business research group studies how co-design, as an approach, can contribute to exploring and addressing complex transition challenges in networks, and thus lead to positive societal change. In this research line, impact means that we empathically pursue social, ecological, and economic values and significance, for individuals, families, teams, neighborhoods, organizations, networks, and our society.

3.   www.intodmentia.nl

Applied Design Research

What is Applied Design Research? The term has a double meaning, as far as I am concerned. For me, design research has always been applied and practice-oriented. It is about research in a realistic context, with people who actually play a part in the societal challenge, and with design (interventions) as a means.

At first sight, the adjective 'applied' seems superfluous. However, to distinguish us from design research at universities, one might say that research at universities of applied sciences has been more applied. Applied in the sense of 'gathering knowledge in, with and as part of the practice'. And bringing that knowledge back to the practice in a form that fits in with that environment.

The latter fits me as a co-design practitioner and design researcher who obtained her PhD as an extra-doctoral student based on the work from own practice. In this social innovation practice, research questions emerged. These questions were encountered by designers, change makers and design coalitions – consisting of stakeholders from government, (non)profit organizations, knowledge institutions, and other stakeholders – in aiming for societal impact. I am now trying to support them by creating, deploying and testing practical, empathic co-design models, methodologies and methods.

149

Societal Impact Design

Collaborating in broad coalitions to achieve societal impact is challenging for many reasons. First of all, it is not easy to 'solve' complex challenges 'just like that' with current innovation instruments. Solutionism is no longer sufficient.[4] Secondly, there are many different stakeholders involved and thus many different perspectives and roles. This makes it difficult to understand and comprehend the situation well, to get an overview, to see alternative futures together, and to arrive at supported and joint decisions and action. Thirdly, societal challenges require a new approach in which citizens, profit and non-profit organizations, and government are willing and able to take on their roles more often and prominently.[5] The research line Societal Impact Design therefore contains three areas of focus on co-design processes: opportunity-oriented design, mixing perspectives, and empathy.

Opportunity oriented design

Opportunity-oriented design assumes that problematic situations cannot always be solved, but we can (learn to) relate to them differently.[6] [7] The responsibility for this lies not only with research, government, profit or non-profit organization(s) or the individual, but is shared by these parties. Issues such as dementia cannot simply be diagnosed or cured with a pill or surgery, unlike, for example, a broken leg (solutionism). They require a more holistic and systemic look, such as a manual therapist looking at your body's entire system to eliminate certain symptoms.

This is also the case for societal challenges. These often require individual and collective behavioral changes, either consciously or unconsciously.[8] Dealing with dementia, for example, is a complex societal challenge for the collective. Many different stakeholders have an essential role when someone has dementia: the person with dementia, their family members, friends, GP, case manager, but also health insurers, healthcare and social institutions, and the local authorities. All of these stakeholders have their own perspective, approach and interests, and see other opportunities for the same issue: how to optimally support the person with dementia and make them enjoy life. The challenge is to approach this integrally.

Mixing perspectives

Secondly, we believe that social innovation and impact can only be achieved by including all the different perspectives of those involved in an challenge, including one's own. As an example, I was the informal carer of my mother-in-law with dementia. This brought me much relevant experience, understanding, and sensitivity during the Dementia Simulator study, but possibly also some bias. That is why the research line is active in studying and explicitly encouraging the mixing of three basic perspectives at every stage of a design process.[9]

In relation to the work and experiences of others, one's own relevant experience can be used more consciously and meaningfully, and possible bias can be prevented in designing with and between others. First of all, it is important to be aware of your own relevant experiences, feelings, but also assumptions and prejudices (first-person perspective). In addition, it is essential to want to hear, see and understand the perspective and experiences of others (second-person perspective). Finally, it is vital to incorporate the knowledge and work of others (experts), such as theory and data, but also documentaries and design of others (third-person perspective). This will provide a good overview of the playing field and the opportunities and dilemmas to achieve positive impact.

Empathy

Next, our Societal Impact Design research focuses on what empathy and empathizing can provide in co-design processes: in engaging with different stakeholders, in creating social innovations and in pursuing positive impact. Developing empathy is an individual process. Although psychologists still disagree on the exact definition, they agree that empathy increases as you consciously focus your attention on self and other(s), as well as that you switch, more deliberately, between affective experiences and cognitive processes.[10]

Empathy is an essential skill in collaboration and co-design, but also a key in generating promising common ideas and social innovation. It is important to use authentic stories,

4. Kees Dorst, "The Core of 'Design Thinking' and Its Application," *Design studies* 32, no. 6 (2011): 521–532, https://doi.org/10.1016/j.destud.2011.07.006.

5. Jung-Joo Lee, Miia Jaatinen, Anna Salmi, Tuuli Mattelmäki, Riitta Smeds, and Mari Holopainen, "Design Choices Framework for Co-creation Projects," *International Journal of Design* 12, no. 2 (2018): 15–31.

6. Caroline Hummels and Joep Frens, "Designing for the Unknown: A Design Process for the Future Generation of Highly Interactive Systems and Products, in *Proceedings of the 10th International Conference on Engineering and Product Design Education* (Barcelona, September 2008): 204–209.

7. Koen van Turnhout, Sabine Craenmehr, Robert Holwerda, Mike Menijn, Jan-Pieter Zwart, René Bakker, "Tradeoffs in Design Research: Development Oriented Triangulation," in *Proceedings of the 27th International BCS Human Computer Interaction Conference* (September 2013).

8. Katja Battarbee, Jane Fulton Suri, and Suzanne Gibbs Howard, "Empathy on the Edge: Scaling and Sustaining a Human-Centered Approach in the Evolving Practice of Design," *Harvard Business Review* (January 01, 2015).

9. Wina Smeenk, Oscar Tomico, and Koen van Turnhout, "A Systematic Analysis of Mixed Perspectives in Empathic Design: Not One Perspective Encompasses All," *International Journal of Design* 10, no. 2 (2016).

10. Justin Hess and Nicholas Fila, "The Development and Growth of Empathy Among Engineering Students," paper presented at 2016 ASEE Annual Conference & Exposition (New Orleans, LA, 2016).

relevant experiences, aspirations, and feelings of people directly affected by an challenge. For example, in the Dementia Simulator project, I developed an Empathic Handover (EH) method for the empathic transfer of my research insights and outcomes to the design team that, for ethical and economic reasons, could not be present in the research itself.[11] Empathy with the design team eventually led to meaningful experiences in the simulator, which in turn led to empathy in its visitors and to behavioral change in daily life.

Experience, interest and expertise

Based on the Empathic Formation (EF) compass,[12] Figure 3 shows an overview of how to look at your own experiences, interests and expertise. The two axes represent the empathic formation process, and the spheres with numbers represent the three basic perspectives discussed above. With this idea, I want to clarify that mixing perspectives and being aware of your experiences, interests, and expertise, and that/those of other stakeholders, is crucial for collaboration. This applies both to managing stakeholder expectations and to realizing societal impact and concrete results.

Figure 3
Stakeholders' experience, interest and expertise.

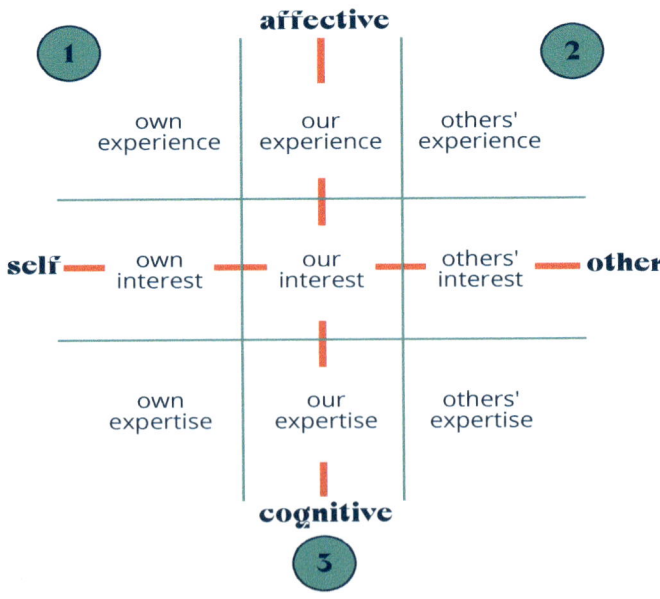

This new idea relates to my previously discussed work and to a recent practical study that I conducted with a shrinking municipality and its inhabitants: the Co-Design Canvas,[13] an empathic co-design instrument with societal impact. The tool explicitly challenges stakeholders to name and discuss everyone's individual and organizational interests, knowledge and power from the start.[14] Particularly, power is a relatively undiscussed concept in the design world.

Conclusion

Societal challenges require a multi-stakeholder approach: holistic, integrating different perspectives and including empathy, and facilitating citizens, (non-)profit organizations, and government to take on their role more often and prominently. By consciously setting up a co-working process in a network that stimulates people to be self-aware and creative, existing structures can be broken up, leading to acceleration, flexibility and more diversity in tackling challenges together.[15]

Working in networks make it necessary to introduce a systemic perspective in addition to empathy. But to be honest, designers are not (yet) trained for that. They have proven to know much about designing for 'gesellschaft' (society), but have not (yet) been properly equipped for designing in 'gemeinschaft' (in networks),[16] which provides for new design opportunities. This requires collaboration with human sciences areas and cooperation with a different, more multi-stakeholder field than designers are used to.

Future work: Systemic co-design

I am pioneering with this new approach and am researching ways to creatively, meaningfully and efficiently collaborate on societal challenges. That is the reason why my fellow professors and I plan setting up an Expertise Network for Systemic co-design with professors of the Utrecht and The Hague University of Applied Sciences. In this network, we can further develop our vision of Systemic co-design (see Figure 4).

11. Wina Smeenk, Janienke Sturm, and Berry Eggen, "Empathic Handover: How Would You Feel? Handing Over Dementia Experiences and Feelings in Empathic Co-Design," *International Journal of CoCreation in Design and the Arts* 14 no. 4 (2018), https://doi.org/10.1080/15710882.2017.1301960.

12. Wina Smeenk, Janienke Sturm, and Berry Eggen, "Comparison of Existing Frameworks Leading to an Empathic Formation Compass for Co-design," *International Journal of Design* 13, no. 3 (2019: 53–68.

13. Wina Smeenk, Anja Köppchen, Gène Bertrand, *Het Co-Design Canvas. Een Empatisch Co-Design Instrument met Maatschappelijke Impact* (SIScode project, 2020).

14. Jung-Joo Lee, Miia Jaatinen, Anna Salmi, Tuuli Mattelmäki, Riitta Smeds, and Mari Holopainen, "Design Choices Framework for Co-creation Projects," *International Journal of Design* 12, no. 2 (2018): 15–31.

15. Jos van den Broek, Isabelle van Elzakker, Timo Maas, Jasper van Deuten, *Voorbij Lokaal Enthousiasme. Lessen voor de Opschaling van Living Labs* (Den Haag: Rthenau Instituut, 2020).

16. Ferdinand Tönnies, *Gemeinschaft und Gesellschaft (1887)* (Whitefish, MT: Literary Licencing, 2014).

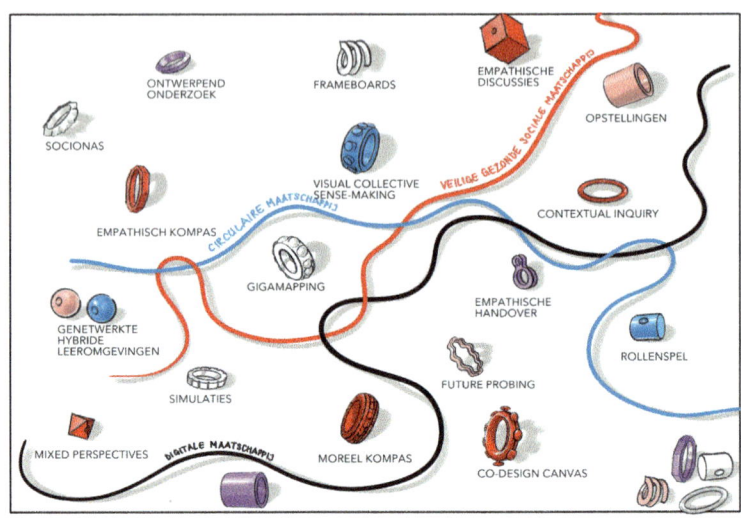

Figure 4
Systemic co-design
bead chain.

Systemic co-design is seen as a multi-stakeholder approach that brings together co-design, the systemic perspective, and human sciences to address wicked societal challenges and herewith accelerate transitions in networks. It is a meaningful and valuable addition to the current set of innovation instruments. Although there are enough Systemic co-design aspects (beads) such as models, methodologies and methods to identify, we need to find out when to use them: in which combination, order and context? How do you 'string' a chain of beads that matches the situation, because no situation is the same (see Figure 5). Some scholars already have experience with using Systemic co-design aspects, and others have thoughts about it, but we need to look at how this can be validated and made accessible to stakeholders *and* even more generic within different transition challenges. In and with our expertise network, we can then realize more meaningful innovations which have broad support and ownership for society, networks, organizations, neighborhoods, teams, families and individuals.

Impact and effect

Finally, my work (models, methodologies, and methods) contributes to further impact in knowledge development, social innovation, personal development, and system development. It also has an impact on society, practice, research and education. The Empathic Handover approach is being used and further developed by a Belgian researcher, who due to the pandemic could have limited contact with people with dementia. The Dementia Simulator will have a follow-up in a Virtual Reality version which makes the experience available to even more people. The Co-Design Canvas has been picked up by our Dutch ministry of Healthcare (VWS), the Pharos expertise center, and various Dutch municipalities as part of the program 'Kansrijke Start' (Promising Start). The tool is also shared with the Mediawijzer network for the benefit of inclusive media design. Even more, the latter has also demonstrated its usefulness for lecturers, students and working field partners within our Inholland urban living labs. In this way, we gradually and jointly accelerate our collective innovation capacity.

17. Wina Smeenk, *Navigating Empathy: Empathic Formation in Co-Design*, PhD Thesis, Eindhoven University of Technology (2 December 2019).

Wina Smeenk

InHolland University of Applied Sciences

Dr.ir. Wina Smeenk studied Industrial Design at the Delft University of Technology. She then worked as an innovation strategist and designer for international companies such as Giant Bicycles, Sony, and PlayStation. After working as a program manager for the Industrial Design course at the Eindhoven University of Technology, she also developed innovative design-oriented educational programs for the Inholland, HAN and HvA universities of applied sciences, and at THNK, the Amsterdam School of Creative Leadership. Since 2010, she has also been running her own co-design agency called 'Wien's Ontwerperschap'. In 2019, she obtained her PhD with the thesis 'Navigating Empathy, empathic formation in co-design processes'.[17] Since 2021, Wina has been a Professor of Societal Impact Design at the Inholland University of Applied Sciences.

155

Designing our society together

With a different language towards a desired future

Christine De Lille

I am a designer, and I do research. Even more, I do design research. That research is also valuable to designers. Are you still with me? I trained as a designer in Delft. The focus of my training was on the people we design for. The methods we used during the design process were often developed in an academic context or within large companies. That is why I chose to focus my PhD research on determining how small businesses could design for and with their users and customers.

My PhD research taught me that small businesses could use the same methods and integrate them into their daily practice. For example, small businesses would not just be having a coffee with a customer but using that moment to gather information and use that information. Small businesses are agile and can respond quickly to this information. When the entrepreneur says this is how things should be, that is what is going to happen. In addition, collaboration offers small businesses many opportunities. This is a true example of how you are stronger together. More about this later.

Design research

In this example, I am studying small businesses. I carry out design research because I collaborate with them; by designing possibilities, I learn what suits small businesses. The knowledge that this provides helps both small businesses and designers who work within and with such companies.

Design in research has many advantages over other research approaches. Almost all types of research are based on the existing situation and study how that situation is handled. That means you always use what is familiar, what is known, as your base. That gets us stuck. Designing enables us to create a new situation and to study that situation. As a result, design researchers can explore new possibilities. They do not use the 'what is' as their base, but rather the 'what can be'. This allows us to get away from the past, learn what might be and take the new possibilities further.

Figure 1
Designing Futures Model, Mission Zero.

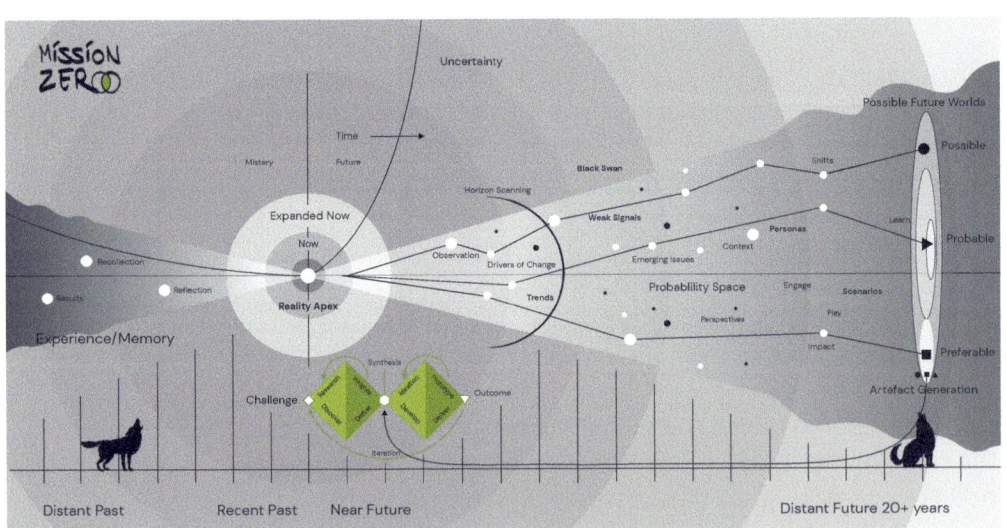

158

Designing a desired sustainable future

1.   Elizabeth Sanders and Pieter Jan Stappers, *Convial Toolbox. Generative Research for the Front End of Design* (Amsterdam: BIS publishers, 2012).

The Innovation Networks research group is also part of the Mission Zero Center of Expertise. In this Center, we want to actively start creating a desired sustainable future. We do not abide by a future that may happen (probable future), or that can happen (possible future). We want to work on a desired future: a preferable future. (See Figure 1 for an overview of the working method, based on the work by Steven Santer.) Through design research, we can consciously use methods to explore a possible future and make conscious choices toward a desired future. Through design research, 'we can study the future by creating possible futures,' and we can see the potential impact of the different future scenarios that have been created. We can move away from a future that may be waiting for us toward a future that we want.

A different language

This 'making possible futures' indicates a different strength of designers and design researchers. They use language differently. Not the language of words, but that of images and form.[1] People say that a picture says more than a thousand words. A picture in itself does not say more, but above all, it says something else. It has nuance, it is less or more explicit, and it also appeals to your intuition. This is illustrated by the drawing by Manu Cornet (Figure 2). We can try to describe organizations, how they are structured, their culture, and the consequences. The drawings show this quickly but powerfully and carry many feelings with them.

The central heart of Apple, around which everything is organized, actually does not differ much from Jeff Bezos who controls everything at Amazon. Yet, the drawings convey a different feeling. Or what about the pistols between the teams at Microsoft? They literally illustrate the killer competition between teams. How does Oracle keep up with a huge legal team and a much smaller engineering team that should actually be the core of the business?

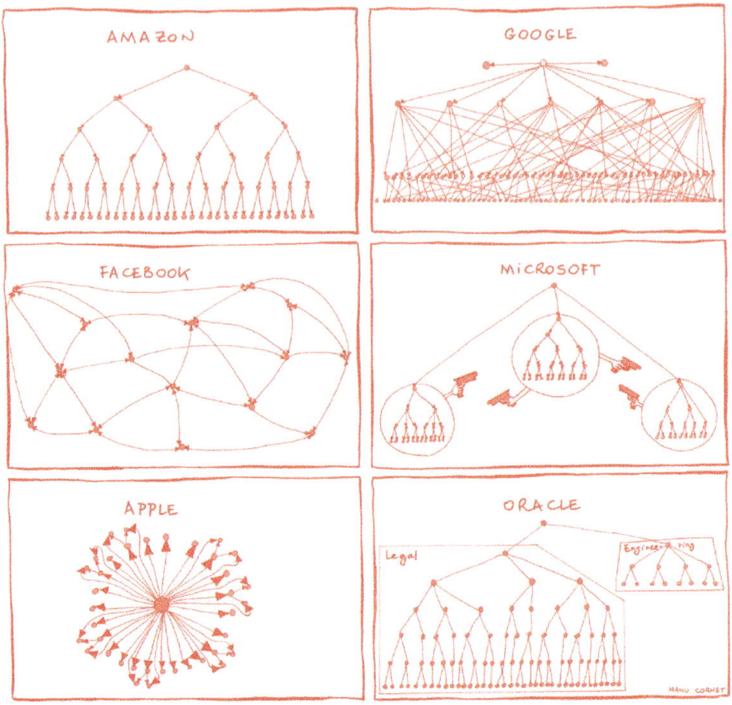

Figure 2
Org Charts, Manu Cornet,
created june 27th 2011,
http://www.bonkersworld.
net/

Working with a different language within research is relatively new. It is often seen as frivolity because it may
look childish and raises the question: how do you capture
something, other than in words? At the same time, it allows
us to design with organizations: how can the organization be
seen as design material? Can we play with this by drawing
structures, like Manu Cornet? And how can we try this out
in real life? These were the questions I also asked myself
after my PhD research. Smaller companies are stronger by
working together. How can we design and explore such
collaborations?

Designing collaboration

Over the last four years, this was the topic of the Innovation
Networks research group. How can we design collaborations? The issues we face as a society, such as energy
transition, sustainability, and security, are no longer attrib-
utable to a single organization. These types of issues can
only be addressed by several organizations together. There
is also no longer a clear client; we all have to deal with the
problem together.

160

There is a role here for the government, as a representative of the people and our society. This government is used to making policies to protect us, but it takes much effort to design, be active, and disconnect from the existing situation. The current issues are often considered complex, difficult, elusive. Through their way of working, designers can offer a grip on this, a language, an opportunity to look at society from a desired vision of the future.

We have to realize that our society has come about through choices in the past. And that new and different choices will allow us to change society to a desired form. Our society can therefore be designed. But how do we do that? Where do we start?

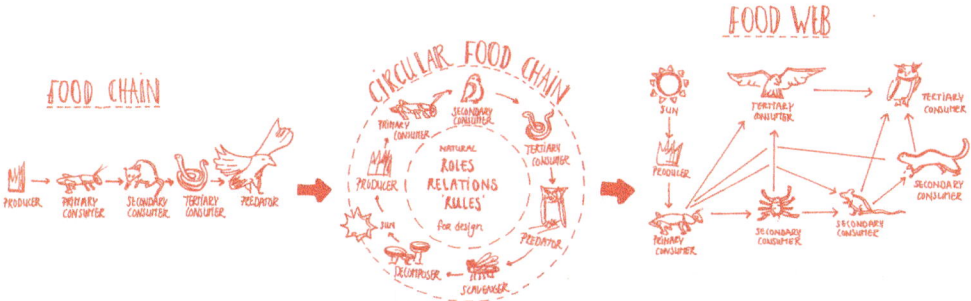

Designing our society, step by step

In her research, Marjanne Cuypers-Henderson looked specifically at the context of our food system. Why don't we look at nature as an example in which collaboration and the whole system are not organized according to food or in a circle, but more like a 'food web', so that if one organization (or in this case, organism) falls away, the web can survive. Figure 3 shows how we might move from a chain to a circular chain to a 'food web'. How can we get organizations to work together, so there is no waste, with one feeding the other, and at the same time not being vulnerable when one of the organizations is eliminated? Currently, Marjanne has her own company, where she looks at how to set up a new network for seaweed by experimenting, trying out, and exploring with others what is possible in the future.

Figure 3
From Food Chain to Circular Food Chain to Food Web by Marjanne Cuypers-Henderson.

161

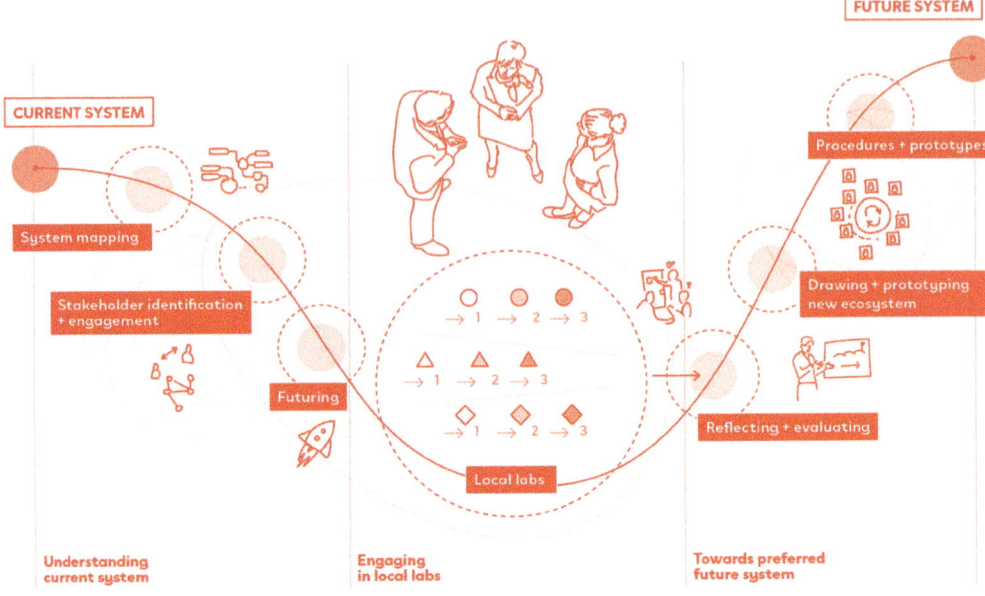

CURRENT SYSTEM

System mapping

Stakeholder identification + engagement

Futuring

FUTURE SYSTEM

Procedures + prototypes

Drawing + prototyping new ecosystem

Reflecting + evaluating

Local labs

Understanding current system

Engaging in local labs

Towards preferred future system

Figure 4
Impacting Systems with Labs. De Lille, C & Overdiek A. 2021 ©

Getting started locally

Another example is that of the Future Proof Retail project. What started with a small project in The Hague, 'From Pop-up to Local Hero' by Anja Overdiek, soon turned into feedback sessions with the National Retail Agenda, at the request of the Municipality of The Hague, to provide input on how we can stimulate the retail sector as a whole. A nationally supported program was set up using previous experience in the aviation industry:[2] Future Proof Retail. The research group was the base for coordinating this program, with more than fifty partners, including fourteen municipalities, trade associations, six universities of applied sciences, and three intermediate vocational education institutes. We used our expertise to design a new collaboration that was scalable and allowed us to take a step back after the program ended.

Crucial in this was to connect the local and national levels. We developed local labs, a total of 22 labs in 2.5 years, using six different formulas. In these labs, local retailers and their employees worked with other retailers and industry associations on themes that enabled them to become future-proof. The experience was shared between the labs and nationally, which allowed the government to support the retail industry on a larger scale.

In this program, we worked as designers to focus on the future, in the context of our research, by designing new possibilities with a different kind of language. We designed and explored not only other forms of collaboration, but also ways to learn from each other. The results of this program are incorporated in the book *Innoveren met labs* (Innovating with labs) by Anja Overdiek and Heleen Geerts,[3] together with the project partners.

Connecting locally with society

In this project, the labs were not isolated activities; they supported the entire industry. This systemic approach allows us to design on a larger scale to impact the overall system. This starts with a good understanding of the current system: what are the issues, and which organizations have an important role to play in this? What are the interrelated dynamics and relationships? And where does the current system run into problems? Why is the desired future not achieved? By mapping out these questions, the system becomes discussable, and we can get a grip on it. In this, we explore the language of the system and its thinking. What are the everyday concerns and power relationships? We can use labs to try this out, in collaboration with others, and experience the possibilities. The lessons from the labs enable us to design a new future system: a desired society, designed according to the process in the model (Figure 4).

With a different language towards a desired future

The labs play a central role in enabling possible changes, although labs are not the only possible form. What is similar about the different forms is how we facilitate a dialog and make the intangible experiential. Designers often work with prototypes, an experiential design, a reflection of their idea, which can solve the problem. The aim is to experiment, to create something similar that can be used.

2. Rebecca Price, Christine De Lille, and Katinka Bergema, "Advancing Industry Through Design: A Longitudinal Case Study of the Aviation Industry," *She Ji: The Journal of Design, Economics, and Innovation* 5, no. 4 (2019): 304–326.

3. Anja Overdiek and Heleen Geerts eds., *Innoveren met Labs, Hoe Doe Je Dat? Ervaringen van Future-Proof Retail* (The Hague: The Hague University of Applied Sciences, 2020).

Figure 5
Together with Being a Designer, the research group developed a 'lid set', allowing people to both experience and change a system.

For example, take the different lab formulas of Future Proof Retail, where we use the different lab formulas to see which works best. To make a system experiential and negotiable, we can also use another type of prototype: Provotypes (Boer, Donovan & Buur). These are 'Provocative Prototypes'; prototypes that want to encourage dialog, create a new framework, and provide a language for it. So, it is not about making the prototype a reality and it being the solution to a problem. Instead, it is a way of making the participants or users change their minds.

One good example of this is the 'lid set', which the research group created for the Mission Zero Center of Expertise (see Figure 5). In the energy transition, much is being said about how we, as a society, can move away from gas, more specifically from natural gas. This is a very current problem in light of the earthquakes in Groningen. Many municipalities inform their inhabitants that they have the ambition to be 'free from natural gas within ten years'. But how do we make this possible? What does the desired future of the energy system look like?

To facilitate this conversation, we have put the most important organizations and drivers on 'lids'. During two workshops, with more than 50 participants, we first mapped out the current system. This went very smoothly. When we asked the participants to shape the desired system with the lids on their tables, everyone stalled. It isn't easy to get away from the existing situation. At almost all tables, the shape of the system was pushed to the side to start over. By using the lids, the system became experiential and at the same time, makeable. A new language had been created, which was shared by all participants. They felt able to talk about complex subjects, point out the mutual relationships, and get a grip on the situation. This is an essential first step towards a desired society.

To achieve a desired society, we must first be aware that there is a difference between a future that happens when we hold on to society as it is now, a possible future, and a desired future. As soon as this awareness is created, we can, by means of design research, create a dialog that allows us to take steps in the desired direction.

Christine De Lille

The Hague University of Applied Sciences

Dr.ir. Christine De Lille is Professor of Innovation Networks. Her research focuses on designing innovation networks, how these can collaborate – i.e., in living labs – and the role of design in research. With her research, she wants to impact at a system level, as she has previously done in the manufacturing, textile, aerospace and retail industries. Christine is co-director of the Mission Zero Knowledge Center. She is also an assistant professor at the Faculty of Industrial Design of Delft University of Technology. She has been involved in several European projects and has extensive experience in leading large consortia.

Integral development of the built environment

Shaping joint responsibility through common understanding

Perica Savanović

Built environment is something we all are familiar with, based on our personal and daily experiences. To speak about and discuss built environment is in a way similar to discussing a national football team. Everybody interested has a sense of expertise, is clearly vocal about it, and would love to have some say in the direction of its (game plan) development. Whereas in the case of football, this complex participation is highly debatable, in case of built environment it is almost entirely justifiable – since we all, as individual citizens and the society as a whole, have to live in and deal with it.

Practice-driven research at Avans University of Applied Sciences therefore approaches built environment from an integral perspective, focussing on participative co-creation as a way for a joint *'onderzoektocht' (exploration)* for professionals from different disciplines and a variety of (end) users. Built environment is literally the place where all the stakeholders come together and, even if often they hold very different 'world views' and perspectives, this offers a great

167

possibility to make the results of collaboration between the public sector, private sector, civil society and academia (often referred to as a 'quadruple innovation helix framework' within a knowledge economy) be felt and tangible.

As a professor in built environment, I professionally bring a background in electrical engineering and architectural design perspective to the table. My formal education started in former Yugoslavia at a broadly oriented electrical engineering professional school, where, in addition to science and engineering subjects, history, literature etc. were also given proper attention. Interrupted by the war at the start of the 90's, I resumed my education in the Netherlands. I quickly encountered a 'cultural shock' of a fairly reductionist technical university engineering education in a predominantly male environment. Looking for a more open, holistic and human-centred approach, instead of largely closed technical systems approaches, I made a switch towards architectural design. Later on, based on examples in practice where I repeatedly witnessed continuous misunderstandings between designers and engineers, the former focussing on synthesis and the latter on analysis as a way to understand an assignment or a given situation, I successfully pursued a PhD on an integral design method (a process of integration of design and engineering disciplines) in the context of sustainable building design (innovations based on integration of human and technical aspects).

Personally, as someone who has experienced what the disintegration of a country and a society means, I bring in the perspective of a European citizen interested in ways of jointly creating a sustainable built environment for the resilient society.

Design, research, design research

My interpretation of the relation between design(ing), research and design research is based on my position as a designer and an engineer that is driven by knowledge development and innovation. Therefore, for me it feels as completely natural that design(ing) and (design) research go hand in hand.

In building practice, design is often, or perhaps even commonly, seen simply as a set of methods to solve problems using existing knowledge and already present (sub)solutions. An approach to match the existing technology, developed 'independently' and elsewhere, with client and/or user requirements. This is partly due to the building procedures that focus on risk management, proven solutions and guarantees. And even though this matching is not bringing us (fast enough) towards a sustainable built environment, this practice generally persists.

Figure 1
Simplified housing renovation project process.

For example, in most renovation projects led by housing corporations, a set of requirements is described; on this basis, the companies from the construction industry are asked to provide a solution (Figure 1, step 1). The contractors then usually gather different professional disciplines to 'solve the problem' and provide a set of feasible possibilities (step 2). The preferred solution is selected, together with the client (step 3). Then a minimum of 70% of the residents/end-users have to agree that the presented solution is acceptable (step 4). The construction project is then implemented (step 5), and at the end, an evaluation takes place (step 6), if at all. For any new initiative regarding the same type of tasks, the same procedure is repeated, even when the same parties are involved.

169

Expected to be primarily a problem-solving activity, it is often overlooked, also by designers themselves, that through design and designing, new and tangible knowledge can be created 'along the way' – allowing innovations and expanding the possibilities by collaborative 'reflection in action'.

Therefore, I think and argue that it is necessary to include the perspective of design research, which explicates this aspect of conscious and deliberate knowledge development by design(ing) in order to give it a deserved place among other ways of knowledge development. For me, this merging of design and research, together with the merging of problem-solving and possibilities creating views, involving various professional disciplines and non-professional stakeholders, resulting in new processes and products in practice, is the promise of applied design research.

Applied Design Research

The (top sector of the) creative industry in The Netherlands and design research at Dutch technical universities generally position design and designers in relation to the national 'Mission-Driven Innovation Policy' from a specific perspective of 'Key Enabling Methodologies'. To connect to a policy based on research views mainly derived from the natural sciences and technology perspective (hence the mirroring of the terminology 'Key Enabling Technologies'), the KEMs are described as design methods to help realize the predefined missions.

This approach is very similar to the current building practice, where designers and design are seen as a problem-solving tool to reach predefined goals, using the existing knowledge and its extrapolations. A very understandable and recognizable approach if the goal is the optimization and reconfiguration of the built environment as a closed technical system. And probably even very efficient in this way, especially in the built environment, where the procedures are already such that the problem definition is initially 'set in stone'. Then the tendering and purchase take over, and in the remaining space, the designers are asked to solve the resulting, often contradictory and/or 'wicked', problems. This approach is highly questionable if we are to make a major leap towards a sustainable built environment.

Design, as a strictly problem-solving activity, and research, as a purely knowledge-developing activity, are this way separated and insufficiently cross-pollinated, in both directions – from research to development and from development to research. Applying the intermediate design results and the design-driven newly acquired knowledge to, for example, reframe a task at hand is usually barely possible because of the linear '(problem) analysis-first, (solution) synthesis-second' based approach.

Applied Design Research can help implement design methodologies as a way to jointly evolve requirements and solutions, demand and supply, questions and answers. Designing, researching and directly applying the newly developed knowledge and innovations to (re)define both the goals and the solution possibilities. Applied Design Research can show, through explicit knowledge development, where and why in the process of designing and in a specific state of design, adjustments in the requirements, criteria and/or general descriptions of the context/situation are welcome – and how different stakeholders can play an active role in this reframing. A design methodology incorporates various aspects of change, including redefining the roles etc., and represents much more than a design method.

It is important to emphasize here, certainly regarding built environment, that it is not only about getting different types of professionals, or different design and engineering disciplines, to work together or better reframe and (re)create their joint understanding – translated and made tangible by designs – but to also actively involve non-professionals (users and citizens, policy makers).

Going back to the renovation project example in housing, an intervention in the process could be based on the initial solution decision and could, for example, involve the end-users in evaluating the renovation of each house or building part (Figure 2, step 2), together with all the professional disciplines (step 3). The collaborative reflection directs the (changes in) further construction (step 4), and the process is iteratively repeated until the whole building is done (Figure 2, step 5). All the improvements along the way can, where possible, also be implemented 'backwards', resulting in a continuous construction improvement project. The end result (step 6) would most probably be markedly different from the choices traditionally made (Figure 1, step 3), but most importantly, it would acquire more than 70% of end-user support along the way. Additionally, this applied design research process would result in a number of extra variations and different (sub)solutions already 'validated' by the end-users could be used as a kick-start for new collaborative development projects (Figure 2, step 1). From a design perspective, construction projects could be steered as a development through direct implementation type of projects.

Applied design research regarding built environment is about a multifaceted knowledge and innovation development that requires embracing complexity and dealing with this complexity using design methodologies.

Within built environment we need to learn how to create new options and possibilities for joint sustainable future(s) together. This creation aspect precedes decision and/or selection-making on which possibilities to further pursue or use. Moreover, it precedes the final definition of criteria for decision/selection making. If applied design research would help change the common view in built environment developments that design is essentially making choices / decisions / selections, and the essential ones exclusively by the 'one that pays' – the impact would be enormous!

Looking from a broader societal perspective, especially in times of bigger polarisations, learning to create new possibilities together is essential to pass on to new generations. Applied design research as a way to fuel this change fits in seemingly perfectly with the Universities of Applied Sciences, since their practice-driven research actively involves students and teachers in an intimate connection between research and education.

Current technocratic trends and future creative challenges

Unfortunately, the latest trends regarding built environment point in a technocratic direction. Where the expectations five years ago based themselves on the further development of the 'golden triangle' of business, government and academia – towards a direct inclusion of users and citizens in 'quadruple helix' innovation processes – we have witnessed an increased focus on bilateral relations grounded in the client-contractor relationships. Technical systems engineering, still perceived as relatively new in built environment, encouraged thinking that we still can decompose and simplify the challenges ahead (even those as big as climate change!) and that we need to continue to educate new generations of professionals to act from within their own discipline silos.

Figure 2
Simplified housing renovation project process.

Design within built environment goes in the direction of standardized solutions, partly because of the imposed interpretations of the industry that a 'platform approach' requires product standardization for mass production. The potential of the still existing craftsmanship in the construction industry, which combined with new technology enables development through mass customization instead of mass standardization, is largely neglected.

173

The main challenge remains to establish learning communities to creatively apply applied design research for built environment; hands-on, for gaining knowledge and understanding each other increasingly better along the way, to create a resilient society. Creating a way, a catalyst, to kick-start diverging development, a variety of changes with converging exchanges.

Future professionals in built environment 'designerly' use technology as a tool to shape sustainable living environments and business models, actively engaging with all the stakeholders in an ever-evolving open society. They are an integral part of, and the driving force behind, an open socio-technical system, not merely a cogwheel in a technological world where the only way forward is to optimize the already predefined closed system.

Applied design research in built environment is primarily about a multi-faceted innovation and knowledge development that requires embracing the (contextual) complexity. Being able to cope with this complexity is made possible through use of design methodologies (and thereby their simultaneous further transformation), which stimulate creativity and collaboration, resulting in requirements transcending creations.

Perica Savanović

Avans University of Applied Sciences

Dr.ir. Perica Savanović is a professor of Built Environment at the Avans University of Applied Sciences. He was born in former Yugoslavia and brought his background in electrical engineering with him to the Netherlands. Perica studied architecture at the Delft University of Technology and, after graduation, went to work at the architectural firm of Van den Broek en Bakema. Fascinated by the difference in thinking between design and engineering, he obtained his doctorate with a thesis in Integral Design at the Eindhoven University of Technology. To apply what he had researched in practice, he then went to work as Program Manager for Integral Collaboration at the Construction Research Foundation, and later on as a Development Manager of Integral Housing at the University of Utrecht. In his practice-driven research, Perica explores the relationship between humans and technology and strives to achieve participating creation through communal conceptualization.

175

PART 4:
BUILDING BRIDGES BETWEEN DISCIPLINES

"Once you see the boundaries of your environment, they are no longer the boundaries of your environment."

~ **Marshall McLuhan**

Smart transitions with design

Making the future tangible for businesses and individuals

Anja Overdiek

'We are not anti-social; we're cybersocial.' This is what one student said in reaction to a frequent reproach about social behavior in digital spaces and the increased loneliness of people due to them exchanging physical interaction for digital contact. What could 'cybersocial' mean when we try to see it not as a defense but as an ambition? Make digital and hybrid spaces more inclusive and democratic? How can design and design research of technology refer and react more to social systems, with their contemporary challenges? The Cybersocial Design research group at the Rotterdam University of Applied Sciences was created in early 2021 to find answers to these questions in collaboration with colleagues from the Creating 010 Research Centre.

Nowadays, digitization is driven by large, international technology companies; it is (more or less) socially supported. However, digitization offers many more opportunities for sustainable, inclusive, and healthy social systems. Think of cities, circular and local production chains, new ways of collaborating and governing. To turn these opportunities into

179

1. Maung Sein, Ola Henfridsson, Sandeep Purao, Matti Rossi, and Rikard Lindgren, "Action Design Research," *MIS Quarterly* 35 (2011), 37–56, https://doi.org/10.2307/23043488.

innovations, we need to make new technologies and their potential transparent and tangible for SMEs and individuals. As a form of Social Design, Cybersocial Design focuses on the joint research and design of digital interactions, making a positive contribution to social transitions.

Applied (design) research as a discovery

As an expert on social movements and the close link between social practice and scientific theory, I had turned my back on the university after my PhD. To me, it was too much like working in an ivory tower. I wanted to gain more hands-on experience, and from there perhaps discover greater heights. Applied research has lured me back to research. The challenge to connect current problems and people's frames of reference to abstract ideas about future possibilities is nowhere more extensive than in this field. Nowhere is there so much room for research based on theory and speculative methods, and at the same time room to build the future in concrete terms together with professional practice. As a political scientist and sociologist, I only started studying design and design research in the last ten years. Applied design research has enriched the qualitative research methods I used until then with a future perspective and systematic abductive thinking. Abduction, thinking in new concepts from an intersubjective but also intuitive and esthetic approach (What feels right? What is nice?), in a time of complex problems and fragmented perceptions and 'bubbles', is particularly valuable in a time like ours.

I experienced the impact applied design research can have in this time when I was heading a recent design-led research project in the retail sector. This was a large project at The Hague University of Applied Sciences, carried out with seven universities of applied sciences and fourteen municipalities. The approach was mixed-methodical, but the basis of everything we did was design. Together with a multi-stakeholder group and facilitated by design researchers, we designed different forms of experimental spaces for shopping areas. We then tested these 'lab formulas' locally with small and medium-sized retailers and their employees in

different shopping areas. The goal was to 'deliver' lasting experimental spaces, which help micro retailers and their employees to take steps towards a digital and sustainable business model. Through action design research,[1] we started comparing designs and began iterative testing in the shopping areas, to develop generic and context-appropriate experimental environments.

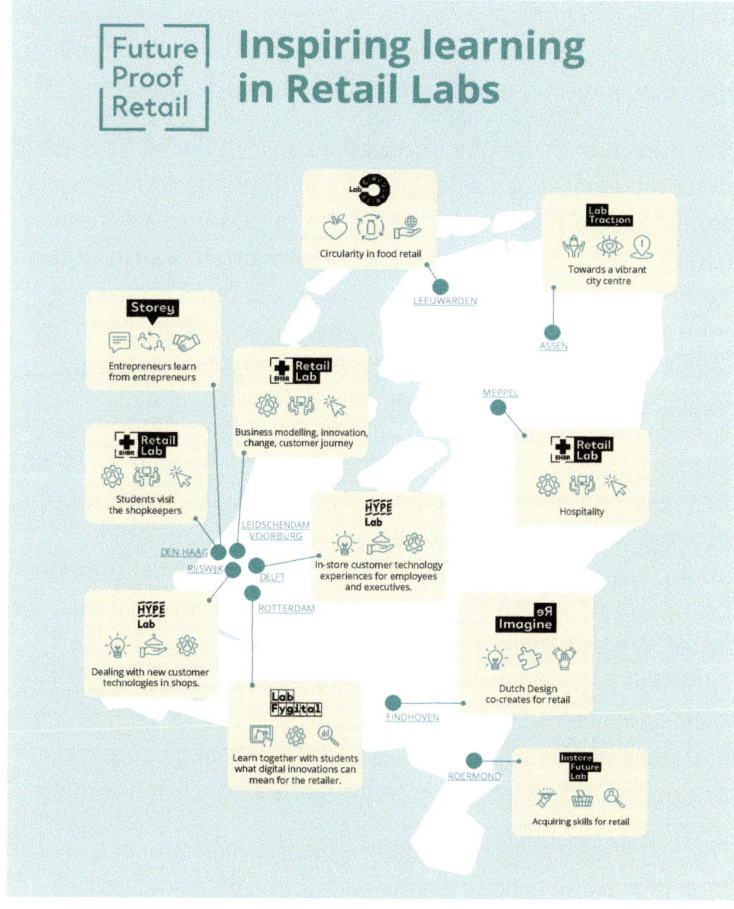

Figure 1
Future-Proof Retail: Eight shopping area 'lab formulas' tested and scaled up over 22 local living labs, © The Hague University of Applied Sciences.

Design innovation from local to system

In the lab formulas designed in co-creation with system stakeholders, we have taken up important societal transition themes one by one: digitization, robotization, sustainability and purpose economy. Micro retailers (1–10 employees), which account for more than 60% of the physical shops in Dutch cities, are not known for their innovative capacity.

2. Elizabeth Sanders and Pieter Jan Stappers, *Convivial Toolbox. Generative Research for the Front End of Design* (Amsterdam: BIS Publishers, 2012).

3. For research through design in the field of human-computer inter-action (HCI), see John Zimmerman, Jodi Forlizzi, and Shelley Evenson, "Research Through Design as a Method for Interaction Design Research in HCI," in *Proceedings of the SIGCHI Conference on Human Factors in Computing Systems* (2007): 493–502, https://doi.org/ 10.1145/1240624.1240704.

4. Anja Overdiek and Heleen Geerts eds., *Innovating With Labs, That's How You Do It! Insights from Future-Proof Retail* (The Hague: The Hague University of Applied Sciences, 2021). Available on www.lulu.com.

5. For systemic design, see *Beyond Net Zero – A Systemic Design Approach* (London: Design Council, April 2021).

6. David Hesmondhalgh, *The Cultural Industries, Third Edition* (London: Sage Publishing, 2013).

However, they have been motivated to start moving by bringing the future to their working environment in a tangible way (Figure 1).

In the local shopping labs, 'make tools' [2] were used to strengthen co-design aimed at forms of collaboration and business models, and to gain insights about new retailer skills. This was more generative and participative research, more research through design. [3] The approach led to many small innovations for the retailers involved and ultimately to the retail industry embracing a 'lab approach to learning'. [4]

The prerequisite for the impact we could make was systemic thinking: exploring the language and thinking, with everyday concerns and power relations in the industry as a system context. We linked this systemic thinking to the local experiments. Future-Proof Retail is an example of an increasing number of similar projects of 'systemic co-design'. With my fellow design professors, I want to explore this relatively new area further in the coming years. [5]

The power of applied design research lies, in my view, in its future-oriented and problem-solving perspective, its merit to provide contextual and participatory customization, its abductive thinking, and its ability to bring together various groups of people through artifacts. I want to explain the latter. What designers (and design researchers) have in common with artists is that they are 'symbol creators'. [6] They create things and services, which give concrete form to ideas and possibilities. Like this, designers can make complex problems understandable and negotiable through 'a thing'.

The Fairphone case is still an impressive example of this. The idea of making mobile phones without rare minerals has long existed. But it wasn't until PR expert Peter van der Mark and designer Bas van Abel created a 'dummy' mobile phone in 2009 and called it Fairphone that all relevant stakeholders from the production chain, governments, start-ups, and consumers joined forces to actually develop, build, market and use this product. This is the typical effect of a 'boundary object' that can bring people from different communities and professional cultures with different 'languages' together for a concrete goal.

In my view, one other good example of problem-solving, contextual, and participative work is the $eev app, created by the 'Afdeling Buitengewone Zaken' design agency, which won a Dutch Design Award for this in 2020 (Figure 2). This app offers a playful solution to youth debt. The app was developed in collaboration with students from the Albeda College in Rotterdam. Also very inspiring is the work of 'future food designer' Chloé Rutzerveld. With the exhibition 'A Radical New Food System for the Post-Anthropocene City' (2020), she lets the general public experience what future food could look and taste like and what effects food choices have on our environment, the body, and the experience of eating (together).

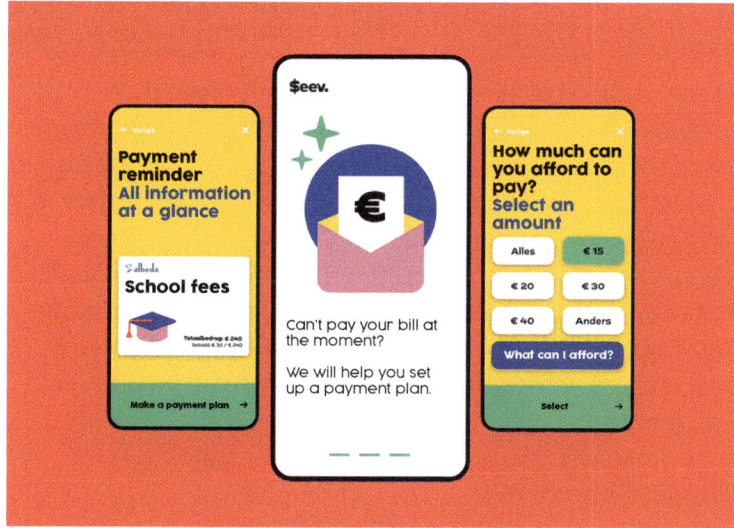

Figure 2
The $eev app for young people with debt problems, © Afdeling Buitengewone Zaken.

Reduce complexity 'by design'

Even closer to home for the digital researcher is the critical work of 'privacy designer' Tijmen Schep, who designed 'Candle – the privacy-friendly smart home', a cloud-free and open-source alternative to the current, data-collecting applications (Figure 3). Candle makes people more sensitive to the privacy effects of conventional smart home devices and demonstrates that there are alternative ways to develop digital technology.

Figure 3
Prototype of the 'priva-cy-friendly smart home' equipment, © Tijmen Schep.

Digital design can also be used in difficult times to strengthen the creative connection between people. That is what the Japanese TeamLab did during the global COVID-19 pandemic. Their online application 'Flower Bombing Home' was accessible via YouTube to anyone with an Internet connection; it used an algorithm that interactively brought together drawings of people from all over the world in one moving 'online painting'. That is how designers and design researchers demonstrate that digitization can be different, i.e., more social. In addition, they reduce the complexity associated with the use of these future technologies for individuals and SMEs.

In the coming decades, the transition to sustainable energy and production systems and to resilient food and health systems will be socially essential. In addition to designing solutions, designers will also increasingly facilitate more social, multi-stakeholder transition processes.[7] Designs (including physical, digital, and mixed artifacts) can make complex subjects manageable. Possible solutions are not only tangible but also testable. In addition to principles for the design of facilitating artifacts, this also requires more systems thinking. The mission of the so-called 'systems-oriented design' (SOD) is to support designers in dealing with complex problems resulting from the increasing connection of people, nature, and technology.

As Birger Sevaldson writes: 'While systems thinking describes the interconnectedness of complex issues, design suggests how to react and innovate as well as solve complex problems. These two modes have not been integrated well enough. The approach of SOD is to build the designer's own interpretation and implementation of systems thinking, so that systems thinking can fully benefit from design thinking and practice and vice versa.' [8]

Complex thinking is challenging, and already developed tools for this, such as giga-mapping,[9] are too. That is why it is crucial that we, as professors in higher education, translate insights and principles into stories and tools, together with and for higher professional education students. Jones,[10] [11] calls this Design 4.0. He has formulated ten principles that connect design thinking with systems thinking. Although recognizing design application areas is essential, I do not think the implicit hierarchy of Jones' model is appropriate, because the design discipline learns across these application areas. Instead, I prefer to base myself on the layout of Price et al.;[12] I consider the different design application areas as concentric circles, with the outer rings drawing from the inner rings' methods and vice versa (Figure 4).

Cybersocial design: A concrete summary

In the coming years, digitization will focus on the development and deployment of data feedback systems on different platforms and the use of augmented or virtual reality and artificial intelligence. As a professor of Applied Sciences, I consider it my role to support SMEs, governments, and individuals with knowledge and practices regarding the social and democratic use of these technologies. The Cybersocial Design research group will do so using two research lines: Smart Transitions and Systemic co-design.

In the Smart Transitions research line, we focus primarily on the much-needed social transition to short and local food chains. In concrete terms, this is about supporting the local/regional network in Rotterdam through interaction

7. Manuela Aguirre, Natalia Agudelo, and Jonathan Romm, "Design Facilitation as Emerging Practice: Analyzing How Designers Support Multi-Stakeholder Co-Creation," *She Ji: The Journal of Design, Economics, and Innovation* 3, no. 3 (2017): 198–209.

8. Birger Sevaldson on www.systemsorienteddesign.net, visited January 2021.

9. Linda Blaasvaer and Birger Sevaldson, "Educational Planning for Systems-Oriented Design: Applying Systemic Relationships to Meta-Mapping of Giga Maps," in *DS 95: Proceedings of the 21st International Conference on Engineering and Product Design Education* (University of Strathclyde, Glasgow, 2019).

10. Peter Jones, "Systemic Design Principles for Complex Social Systems," *Social Systems and Design* (Springer: Tokyo, 2014): 91–128.

11. Peter Jones, "Contexts of Co-Creation: Designing with System Stakeholders," *Systemic Design* (Springer: Tokyo, 2018): 3–52.

12. Rebecca Anne Price, Christine De Lille, and Katinka Bergema, "Advancing Industry Through Design: A Longitudinal Case Study of the Aviation Industry," *She Ji: The Journal of Design, Economics, and Innovation* 5, no. 4 (2019): 304–326.

design and design facilitation. Networks, not individuals or individual companies, will form the base for new ways of producing and consuming food. How can digital technologies strengthen networking for sustainable and resilient new systems? Which data feedback algorithms promote the interests of these transition networks?

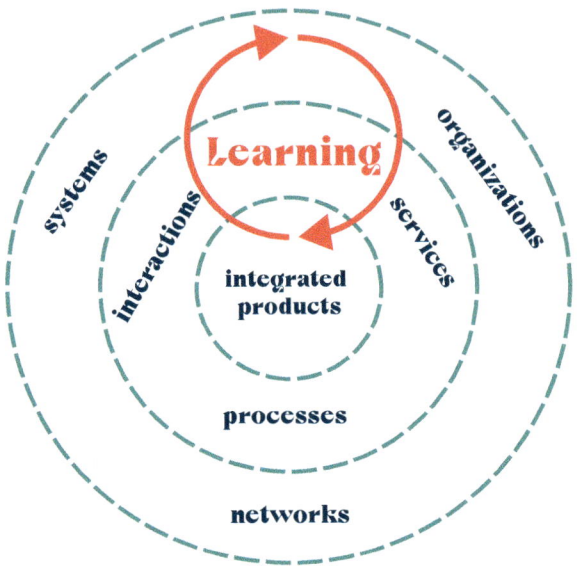

Figure 4
Development of design discipline covering three application areas, based on Jones (2018) and Price et al. (2019) © Overdiek 2021.

What are concrete designs for this, when and with which tools can the local network be supported and scaled up? These questions will be answered in collaboration with regional companies, the municipality, national networks and consumers, and lecturers and students of the Institute of Communication, Media and Information Technology of the Rotterdam University of Applied Sciences. Resulting platforms and insights fit well with the city of Rotterdam and its ambition to be a trendsetter as a 'smart and social city'.

The second research line, Systemic co-design, focuses on the design principles, competencies and tools that designers need to create more systems-oriented designs and to design with various stakeholders. Which competencies are essential for guiding multi-stakeholder networks, and which tools can be developed for this? These include the further development of integrative 'design future' approaches [13] or creating and facilitating using 'cultural probes' [14] [15] aimed at various stakeholder and user groups. It is also important to find out which design principles, but also which specific learning spaces,[16] support design students in higher professional education to develop these skills. In addition to the partners mentioned in the first research line, the research group will also work closely with regional design offices.

Using collaboration and new insights around these two research lines, the Cybersocial Design research group aims to contribute to making the future smart and social by keeping it tangible and influenceable for individuals and businesses.

Anja Overdiek

Rotterdam University of Applied Sciences

Dr. Anja Overdiek has been a professor of Cybersocial Design at the Creating 010 Research Centre of the Rotterdam University of Applied Sciences since January 2021. Her research areas are stakeholder engagement, co-design, and experimental spaces, mainly aimed at societal transitions. She is particularly interested in the opportunities for digitization and interaction design for these transitions. Anja previously worked as an international manager in the IT industry. She was the director of her own organizational coaching company called Human2Organisation. Currently, she is also an associate professor at the Mission Zero knowledge center at The Hague University of Applied Sciences. Anja obtained her PhD in political sciences at the Freie Universität Berlin with a thesis in the sociology of knowledge field. Originally from Germany, she has been living in the Netherlands for more than 20 years.

13. A recent example of such research is Fernando Galdon, Ashley Hall, and Laura Ferrarello, "Futuring and Trust; a Prospective Approach to Designing Trusted Futures via a Comparative Study Among Design Future Models," *Proceedings of DCS conference, Scenarios, Speculation Strategies* (November 2020).

14. For the use of HCI probes, see Kirsten Boehner, Janet Vertesi, Phoebe Sengers, and Paul Dourish, "How HCI Interprets the Probes," *Proceedings of the SIGCHI Conference on Human Factors in Computing Systems* (2007): 1077–1086.

15. Kirsten Boehner, William Gaver, and Andy Boucher, "Probes," in Celia Lury and Nina Wakeman eds., *Inventive Methods: The Happening of the Social* (New York: Routledge, 2012): 185–201.

16. About the use of temporary space as a lab in a real-life context see also Anja Overdiek and Gary Warnaby, "Co-Creation and Co-Design in Pop-Up Stores: The Intersection of Marketing and Design Research?," *Creativity and Innovation Management 29* (2020): 63–74.

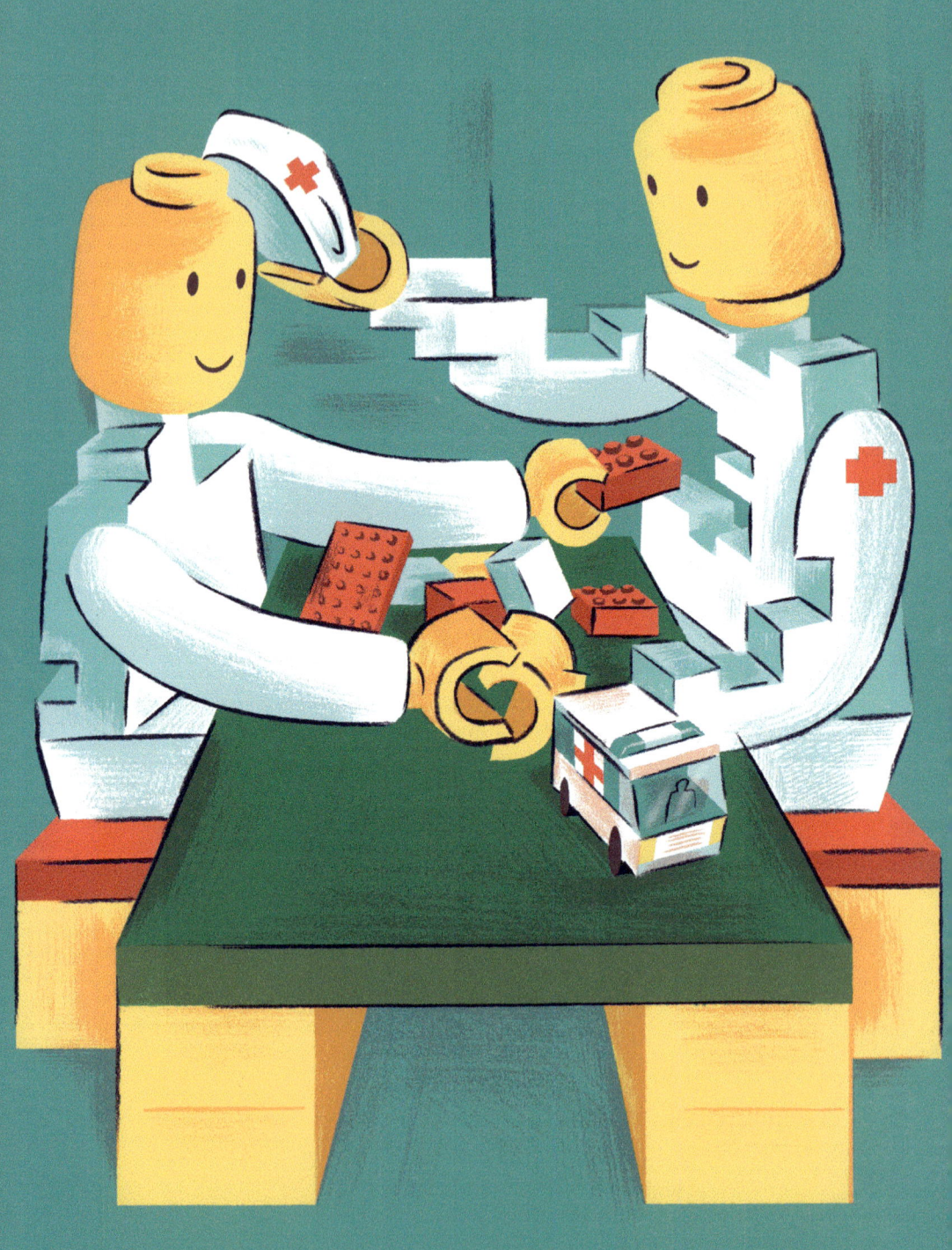

A new mindset in research

Creativity, empathy, and participation lead to successful technological innovations in healthcare

Eveline Wouters

In the NADR network, I am a bit of an odd duck: I am not a designer but have a medical background. Moreover, the term *design* is not mentioned in our research group's name, Health Innovations and Technology, as with most other research groups. But one sure thing is that the term design suits me and what we do in our research.

Over the years, first as a (clinical) researcher, and later as a professor of 'Health Innovations and Technology' at Fontys University of Applied Sciences and as professor of 'Successful Technological Innovations in Healthcare' at Tranzo, Tilburg University, I have gradually familiarized myself with an ever-broadening range of research methods. Design research, which for me is the relatively latest methodology, plays a very special role.

Involve users in healthcare innovations

Our research focuses on using technology in long-term healthcare, focusing on human factors that promote and hinder it. The acceptance by stakeholders (patients, family, and care providers), the implementation aspects in a care organization, and their accessibility and applicability for various target groups and in different circumstances are topics of our knowledge development.

In the recent past, the use of technology in healthcare was often based on a product that had already been designed. It looked at existing technology, which is applicable in healthcare, and then at factors that affect its acceptance. However, the design process itself is strongly and inextricably linked to its subsequent acceptance. The deployment process (implementation) at the back-end cannot be separated from the front-end design process.

That is how design research logically became an exciting and vital subject for us. The involvement of users, taking various necessary steps in the process, and the sound (scientific) basis of these steps, also makes design research a practical framework to naturally link research to the practical environment.

A new mindset in healthcare research

After learning all this, to me, applied design research means more than just a research method. First and foremost, it is a mindset that focuses on the full involvement of end-users in the design of a product, service, or organizational change. It is a form of true democracy, in which the people for whom a design is intended are co-shaping it. Within chronic healthcare, these are the patients, their family, informal carers, and healthcare professionals.

Secondly, it is not one type of research, but a combination of methods that, based on the stakeholders mentioned, can take many forms. Thirdly, applied design research is a way for me to be creative in research without affecting the validity of that research. It is a way to gain more room for solutions other than the usual methods offer. It is also a way to become more forward-looking and solution-oriented and not just evaluating.

Focus on the future of healthcare

Applied design research provides room for deepening its application. That is precisely the element that appeals to me in this research: being creative, going deeper, and the genuine involvement of people who want to create better healthcare. As a researcher, you cannot distance yourself from the participants as you can in quantitative clinical research. It is truly 'applied', in the sense that it takes place in and with the practical environment.

There is also the possibility of trying out a lot, thinking outside the box, and actually coming up with surprising results. As a result, sustainable solutions are created, and we (the health-care professionals) stop repeating ourselves. Also noteworthy is that it is a fun way of working. Because of the creative element, the many possibilities that address both cognitive and emotional aspects, and the participation in sessions by stakeholders, is usually perceived as fun. People get excited and become more creative. The method itself is already enriching, apart from the outcome. I have also found that the participants' involvement and full participation in a design process is generally greater than in 'classical' research.

I have also experienced that applied design research, through its broad range of approach options, offers practical tools to support an impact project pathway. For larger projects, where besides the scientific impact, the social impact is especially relevant, sub-steps can apply design research methods. In an impact project, based on higher targets, the work is done backward towards concrete activities that have an impact. Design methods and the mindset associated with applied design research can support the setting of goals and the iden-tification of desired effects, and the 'measuring' of them.

The essential step towards the future

For me, applied design research means more than what I wrote here. In my already quite long working life (over 40 years) in various branches of science, I have often missed something. I have been classically trained in medicine and epidemiology, with randomized controlled trials, experiments and cohort research as leading research methods. This result is a high degree of evidence, valid for large groups. The development of healthcare has greatly benefitted from this.

While very much appreciating the value of these methods, I always felt something was missing, as I said. Traditional research starts at the group level, is based on hypotheses, and takes place in the present or the past. It generates generalizable knowledge and is therefore of great importance. However, what is lacking is the 'improvement step', the step towards the future. After establishing correlations and, if possible, causal relations, there is still a step that is not so much looking at the present and the past, but wants to improve the future. This requires something else, a different mindset and a different approach.

Figure 1 (left)
Nursing home professionals using LEGO Serious Play.

Figure 2 (right)
Results of the LEGO Serious Play workshop.

A research method with empathy

A second aspect that I missed was the relationship with the practice environment. For example, take the relationship of the discussion between the unique individual patient and the physician, and the scientific facts at group level. For me, applied design research adds something significant to this gap.

What is important to me personally in this respect is that the first step in applied design research, when it comes to participatory design, is 'empathy'. I have found that by using this type of research more often, making it better known in wider scientific circles, researchers who perform other types of research gradually become 'infected' with this consciousness. They become aware that research (in my case in healthcare) is based on knowing and understanding the needs of the care recipient. This relationship between researcher and participant has significant implicit consequences. It means that participants more often will be treated in a different, more empathetic way and will be involved in the research.

1. SCAMPER: acronym for Substitute, Combine, Adapt, Magnify or Modify, Put to another use, Eliminate, Reverse or Rearrange.

2. Eveline Wouters and Joost van Hoof, "Professionals' Views of the Sense of Home In Nursing Homes: Findings From LEGO SERIOUS PLAY Workshops" *Gerontechnology* 16 (2017): 218–223, https://doi.org/10.4017/gt.2017.16.4.003.00.

How it works: Two examples

I want to illustrate the pleasant involvement of participants, the encouragement of creativity, and the finding of solutions with two examples. The first example is 'feeling at home in the nursing home', where we used LEGO Serious Play (LSP). The second example addresses the research ethics approach within higher professional education, where we have applied the SCAMPER method.[1]

Explore the 'feeling at home' with LEGO Serious Play

One of our research programs dealt with developing a nursing home environment, focusing on making residents 'feel at home'. At the start of this program, we wanted stakeholders who work in a nursing home to express what they consider 'feeling at home'. During an evening session, we asked seventy professionals for their opinion.

After a brief introduction, they started working in smaller groups using LEGO (Figure 1). By first building and then naming what they had built, we also appealed to their feeling, their intuition. Then, by letting everyone tell the story of the building, all participants could tell their story and what was told, was remembered better: the image strengthened the story (Figure 2). Although this evening was meant to introduce the 'feeling at home' program, it provided us with such a wealth of information that we wrote a scientific article about it.[2]

Ethical antennas with SCAMPER

Since 2012, the Fontys University of Applied Sciences has had a Research Ethics Committee. This committee assesses human-involved research and advises on the ethical aspects of research to researchers. One of the tasks we set ourselves was to write a manual for researchers. The main objective was to develop 'ethical antennas' in students conducting research.

To this end, we organized two sessions with research supervisors, lecturers, and students. The central question to the second session was 'how do you, as a lecturer-coach, develop ethical antennas in students in the context of their research?' We started using the SCAMPER method. Step 1 was to get the problem clearly in view. Step 2 was to come up with a solution and name the pitfalls in words and drawings. Step 3 was to choose one of the seven methods of SCAMPER (for example, to *ADAPT* the preferred solution or to *COMBINE* elements). On the different work tables, cards with letters were laid out for the participants. The session generated great, useful ideas, which we have incorporated in the book *Ethiek van praktijkgericht onderzoek; Zonder ethiek is het al moeilijk genoeg*.[3] (Applied research ethics; Without ethics it is already hard enough)

What has changed

Looking back on the last five years, I have noticed that the research has developed from almost nothing being done within the context of healthcare and well-being, to design methods being frequently used in research. Design research has acquired an important place in various institutes within the Fontys University of Applied Sciences, next to the more classical research methods. I also see that design thinking is being applied not only in a research setting, but also in other practices, for example to involve more employees and students, but also external stakeholders, from the bottom up, in policymaking. That is a good development.

How will it proceed?

I anticipate that this trend will continue and that design methods will be increasingly applied in a broad field, not only for healthcare itself but also for health education and various services. As a result, those involved have a stronger voice in what the future (of healthcare) will look like. I expect and hope that this will also lead to a mind shift in healthcare and education itself. The fact that people are thinking more out of the box means that the user's voice (whether this is the student, the patient, or the client in a healthcare situation) is heard. As a result of these developments, the quality of healthcare and education is improved; it has become more diverse, more personal, more enjoyable, and better accessible to more people.

3. Eveline Wouters and Sil Aarts, *Ethiek van Praktijkgericht Onderzoek: Zonder Ethiek is het al Moeilijk Genoeg.* (Houten: Bohn Stafleu van Loghum, 2017).

Eveline Wouters

Fontys University of Applied Sciences

Prof. dr. Eveline Wouters is a Professor of Health Innovations and Technology (HIT) at the Faculty of the Fontys University of Applied Sciences, School for Allied Health Professions. She trained as a physician and epidemiologist. In addition to her work at Fontys, she works as a professor of the 'Successful Technological Innovations in Healthcare' chair at Tranzo, TSB, Tilburg University, the academic collaborative center 'Technological and Social Innovations for Mental Health'. In her research, Eveline focuses on the human aspects of technological innovations in long-term healthcare; she seeks answers to how and why technology is or isn't adopted at an individual level, and how technology affects the work and the collaboration of and between people in long-term healthcare.

Focus on the practical question

The meaningful application of technology in healthcare and welfare

Job van 't Veer

In recent years, many healthcare and welfare organizations have experimented with eHealth applications such as electronic patient records, wearables, apps, games, virtual reality, or social robots. In some cases, the extent to which these investments have led to actual use by professionals, clients, or patients is questionable. Many projects that focus on digital (healthcare) applications demonstrate that successful innovation is primarily *not* technical but social.

Within the Digital Innovation in Healthcare and Welfare research group we focus on how technology can be used meaningfully in healthcare or welfare practice. Human-oriented design is, therefore, an essential theme within the research group. Not only to create meaningful digital applications but also to enable the integration of these applications into the practical context. The design process of innovative (healthcare) interventions must consider the social, professional and organizational aspects.

This means that, from the start, professionals and their target groups are required to play an active role in the development of an application and the sustainable integration of this innovation in daily practice. We believe that this gives substance to the 'A' in applied design research. The importance of designing solutions based on the wishes and characteristics of the user is something that is often advocated in healthcare; but at the same time, it is often not consistently followed through. Our research group is focused on developing a vision of what an (applied) design-oriented approach to innovation in healthcare and welfare involves, and how this translates into a sound methodological approach.

Below we would first like to discuss how applied design research has developed in recent years within the activities of the research group both in research projects and in education. Next, we offer a brief look at what could be relevant directions for the further development of applied design research in healthcare and welfare.

Applied design research in the field of healthcare and welfare

In recent years, applied research has been given a more prominent place in higher professional education. Still, it has not been able to distinguish itself much from academic institutions. Although research conducted at universities of applied sciences clearly puts more emphasis on how results and insights are relevant in daily practice, more than is the case in scientific research, much of the research still relies on traditional research methods.

This often translates into interviews, questionnaires, observations, or focus groups for healthcare and welfare. The increased attention design-oriented research received over the past five years, creates an opportunity for universities of applied sciences of to develop a more distinctive and recognizable profile on applied research.

Applied design research in research practice

In applied research, the most suitable methodological approach is determined by how a research question emerges from the working field. Although this can still lead to a traditional research approach (i.e., an exploratory or evaluative study), things are shifting.

Within the Digital Innovation in Healthcare and Welfare research group we are increasingly looking for the potential *design issue* that is bothering the practical environment. The aim is not to apply a design-oriented methodology at all costs, but to explore the problem (or wish) from a design-oriented perspective. Our experience shows that these discussions with professionals in the field often result in widening the scope of one's problems. This frequently leads to the (self)-insight that professionals are inclined to reason from their pre-existing assumptions about (the cause and effect of) an issue. And they disallow themselves to go beyond these familiar frameworks to reach innovative solutions.

By articulating the question through a design-oriented mindset, we, in collaboration with the working field, are more likely to come up with the formulation of projects that: (1) are more focused on the ultimately desired situation, preferably from the perception of the main target groups and (2) are formulated in such an 'open' way that enables a broader view of possible solutions. These essentially different conversations, which we conduct from the start of a project, are essential to stick to a design-oriented approach in the next steps.

MEE Lab: Design-oriented working as a basis for an intensive collaboration with the practical environment

Recently, NHL Stenden and social welfare institution MEE Noord (an organization that supports people with a mild intellectual disability, autism, or other mental disorders) entered into a

199

1. Vijay Kumar, *101 Design Methods: A Structured Approach for Driving Innovation in Your Company* (Hoboken: John Wiley and Sons, 2012).

2. Robert Curedale, *Design Thinking: Process and Methods* (Topanga: Design Community College, 2016).

multi-annual collaboration in the form of the innovation lab MEElab (meelab.nl). The supporting principle behind MEElab is that we explicitly initiate all projects from a people-oriented approach. The MEElab aims to tackle the more complex issues involved when helping the various target groups. One of the first projects focuses on people with mild intellectual disabilities who have committed a criminal offense. These people often commit multiple crimes, and the recidivism rate in this target group is remarkably high. This ineffectiveness does not only appear to be caused by the methodical approaches used by professionals. The processes within the system of collaborating organizations are also far from optimized. As a result, you have a nice wicked problem that the MEElab wants to address, with the participation of the target group itself and the relevant organizations (e.g., the welfare and justice departments, rehabilitation facilities).

Applied design research in education

In recent years, much attention has been paid to the 'research skills and mindset' of students in higher professional education. Parallel to the development of research groups, education has long been based on existing research traditions in the social sciences and medical world. In recent times however, this research vision seems to shift towards a different, more design-oriented interpretation. Working with design methods that stress the importance of empathizing and co-creating with the target group, appeals to the students in the field of healthcare and social work. These forms of work fit seamlessly into the professional framework in which students are trained: they gain insight into the person that requires certain medical or social support and try to create an appropriate care path.

Within the curricula of the healthcare and welfare education programs of many universities of applied sciences, this development toward a more design-oriented approach is becoming more and more substantial. What started in NHL Stenden with some of the Master's programs (Health

Innovation, Design-driven Innovation, Serious Gaming), is now broadening into the Bachelor's programs. In the health-care and welfare programs of NHL Stenden, all students now choose to graduate with a design-oriented thesis unless the nature of the assignment explicitly requires something else. This implies that these design research skills require attention earlier on in the curriculum. And this is developing quickly: the many field labs or innovation hubs, which are set up in conjunction with the practical environment, provide bachelor students in all years of the curriculum the context they need to work on authentic challenges in a design-oriented manner.

Figure 1
The book *Designing for Healthcare and Welfare* [3]

These developments in higher education have also increased the desire for appropriate educational material. Although there are already many (digital) sources about design thinking and the methods to be used within this concept,[1] [2] these often do not self-evidently fit the higher professional education standards. They do not match in terms of depth, and also not in the extent to which the design-oriented ideas are explained to the field of healthcare and social work. This changed in 2020 due to the arrival of a new textbook, which we developed with, among others, Eveline Wouters and Remko van der Lugt. They each describe their experiences elsewhere in this publication.

3. Job van 't Veer, Eveline Wouters, Remko van der Lugt, Monica Veeger, *Ontwerpen Voor Zorg en Welzijn* (Bussum: Coutinho, 2020).

Textbook design-oriented work for (higher professional) healthcare and welfare education

Based on the projects in recent years, the different research groups have gathered the necessary knowledge and experience about a design-oriented approach to healthcare and welfare. In recent years, several universities have been orientating on alternative approaches to applied research within their healthcare and/or social work curricula. For example, to enable students to work more towards relevant professional products in graduation projects and minor assignments. To produce something more than just a research report; something that makes a more tangible contribution to, for example, the improvement of a healthcare intervention or support process.

To this aim, three research groups got together and created the Ontwerpen voor Zorg en Welzijn *book (Design for Healthcare and Welfare).[3] In the book, we explain how a design-oriented approach is relevant to the field of healthcare and social welfare. The book also has a methods toolbox. The explanation of the methods in this book is deliberately more extensive than in other design books. This enables the reader to apply the appropriate work methods methodically. It is estimated that the book is currently used nation-wide at approximately 15 to 20 higher professional education courses.*

Developments in the field of healthcare and welfare

In healthcare and welfare, the professional works with target groups with different characteristics, vulnerabilities, and disabilities, e.g., people with dementia, intellectual, physical or visual disabilities, or people with an immigration background. To come up with suitable solutions for these target groups, the design *process* must also be sufficiently accessible to them. *Inclusive design,* therefore, also requires inclusive design *methods*.

Figure 2
Working with clay during
the online context mapping
session.

The wide variety of design methods potentially offers many
opportunities to improve the participation of people with
disabilities in the design process. For example: methods that
rely heavily on participants' verbal abilities are not always
suitable. Sometimes, more visual or active working forms
(such as photo elicitation, cultural probes or LEGO Serious
Play) are better suited because participants are very young
or have a cognitive disability.

However, there is still much to learn about how these
methods can be more specifically tailored to the various
target groups, so that these participants can demonstrate
their interests and perspective (even) better. The added
value of a (people-oriented) design approach could be
increased by using this methodological refinement in
the coming years. Current projects, as described in the
framework texts, can hopefully act as a model for further
(methodological) development.

Further development of design-oriented methods towards specific target groups

*Which psychosocial challenges do people with a visual disa-
bility face due to the pandemic? And what kind of appropriate
intervention can be designed together with them to better deal
with this? This is the subject of a ZonMw project, led by Saxion*

University of Applied Sciences' Brain & Technology research group in collaboration with NHL Stenden University of Applied Sciences and practical partners Koninklijke Visio, Bartimeus, and the Robert Coppes Foundation. In this project, we have the particular challenge of making the design methods suitable for a blind/visually impaired panel (of which some are also hearing-impaired). Because many design methods aim to make things visual, many of the regular design methods are rendered useless.

It is common, during a so-called context mapping session, to have people draw or build something with LEGO. This should support them in expressing their thoughts and feelings on a particular theme. The objects that are created then provide the direction for a subsequent discussion. These visual options were obviously useless in this project. That is why it was decided to work with clay. Relying on their sense of touch, the participants proved very capable of creating attributes that were supportive of the further discussion.

In parallel with this ZonMw project, Koninklijke Visio has launched a project to develop a more expansive repertoire of design methods suitable for people with visual disabilities. This fits in with the policy of involving the target group (even) more in the innovative projects that will be initiated in the future.

What does this mean for education in the field of healthcare and welfare?

When design-oriented research is given a structural place in the curricula, it will also require further development in terms of the mindset and methodological skills that go with it. This applies to the students, but certainly also to the lecturers who are responsible for the supervision. This will require further investments. Research groups can play a role in the professionalization of lecturers in this area.

The expectation is that increasingly more healthcare and welfare training courses will choose design research as an option for the graduation phase. This makes applied design research a vital element for a program's final level. If this trend continues, programs will have to formulate a clear policy in this regard. This also makes national alignment desirable. What is the added value for career professionals? What is a 'design-oriented professional product' that logically connects with the final qualifications of healthcare or welfare programs? In this respect, attention may be less focused on the methodological carefulness and transparency (i.e., *the rigor*) of research activities. And perhaps more on how the process has led to a well-founded prototype and/or triggered a specific movement among key stakeholders in the practical environment.

In short, in the coming years, there are exciting challenges to further consolidate and develop applied design research in the research and education of higher professional education, including in the field of healthcare and well-being.

Job
van 't Veer
NHL Stenden University
of Applied Sciences

Dr. Job van 't Veer is Professor of Digital Innovation in Healthcare and Social Work at the NHL Stenden University of Applied Sciences. Since 2004, he has been a lecturer for various healthcare and welfare programs, particularly at the Master Health Innovation program. As a researcher, he is always focused on the social participation of vulnerable groups, such as people with mental illness, mild intellectual disorders, and dementia. Since 2012, he has been focusing on digital innovation in healthcare and social welfare. Within all research projects, the emphasis is on a design-oriented approach: how can you, together with clients, residents, and professionals, innovatively and possibly digitally, improve healthcare and social welfare?

205

Shaping an empathic living environment

Design research as an incentive and medicine for healthy living

Masi Mohammadi

Technological and social phenomena require a periodic reorientation of the design of our living environment. How we experience our living environment and the effect of space on our (social, physical, and mental) health cannot be understated. The Chair Architecture in Health – part of the Built Environment Academy at the Arnhem and Nijmegen University of Applied Sciences (HAN) – combines advanced technologies such as Artificial Intelligence (AI) and the Internet of Things (IoT) with social issues such as aging and civil society to shape healthy and stimulating living environments. To that end, the Chair uses approaches such as Active & Healthy Living and considers people's capabilities and conditions. Our ambition is to develop, for, and together with the (end) users, methods, and strategies for healthy living in our increasingly smarter homes and neighborhoods and to put these strategies into practice.

People-oriented approach

The guiding principle of our research is a human-centered approach. By conducting research together with and for the user, our research group aims to develop an empathic living environment. By creating a balance between the

207

inhabitant's needs, housing conditions, and technology, such a living environment enables people to shape their everyday lives (self-reliantly) based on their capabilities and preferences. Although this caring and interactive environment goes beyond today's (im)possibilities and is, in essence, future-proof, it revolves around the 'new (independent) living,' in which the smart living environment helps to keep users physically, mentally, and socially healthy and, where possible, encourages them to a healthier lifestyle. As such, architecture goes beyond assisted living and becomes both a stimulus and a 'medicine.'

My experience in working with industrial partners, civil society organizations, and SMEs, combined with my academic background in several fields (cartography, civil engineering, health architecture, smart building technology), have led to an interdisciplinary and holistic line of research. This is translated into our interdisciplinary research program 'Empathic Environment', where we develop an evidence-based framework for a human-centered embedding of emerging technologies in spatial concepts and systems. To arrive at this framework, real-life experiments in field labs are a prerequisite. In 2017, together with partners from science, industry, and housing and care practice, I set up the Dutch Empathic Environment Living (DEEL) labs to develop and empirically evaluate emerging housing typologies and their impact on users' well-being and social engagement.

Building for a healthy and inclusive living environment

Our research program has three focus areas: Smart Assistive Homes, Architecture of Cohesion, and Sustainable Healing Environment.

Smart Assistive Homes research focuses on innovative architectural solutions for special target groups, such as seniors (with dementia) or people with disabilities. The emphasis is on designing and evaluating smart, user-oriented homes and districts to stimulate health and (independent) living. We developed an Empathic Design Framework to provide the theoretical basis for cross-disciplinary research projects in this ethically sensitive field.

208

The Architecture of Cohesion research theme aims at studying socio-spatial interventions in the living environment to stimulate social health and inclusion. It entails research into (design) guidelines for (emerging) housing typologies as an answer to societal issues like healthy lifestyles, independent living, or regional shrinkage. The social intervention is expressed in providing and implementing mechanisms for residents' empowerment and participation in the design and realization process. This also includes tools for integral collaboration between stakeholders.

Our third research theme, *Sustainable Healing Environments*, focuses on the relationship between sustainability and health. It examines the preconditions and concepts for designing a both healthy and green built environment. This line of our research program builds on existing knowledge in the field of smart biophilic design with a specific focus on the (end) user. Research into biophilic design has demonstrated positive effects on reducing stress, improving cognitive functioning, and enhancing the healing rate of certain conditions. By combining human- and nature-friendly models, this theme explores opportunities to design therapeutic and healing living environments.

Figure 1
An example of our participatory research is the 'Empathic Home,' built in collaboration with over 30 companies and organizations.

Types of research

The Chair of Architecture in Health is primarily concerned with setting up and carrying out high-quality research, gaining and sharing insight into the healthy living environment, and developing architectural and building-technological concepts and instruments to create and sustain such environments. Our research group conducts (design) research on three levels: *academic, practice-based, and applied.*

The academic research is characterized by investigating the theoretical underpinnings of health architecture, close collaboration with other universities in several PhD trajectories, supported by peer reviews (also among our multidisciplinary research group), and output in terms of scientific articles.

Our practice-based (design) research is cross-disciplinary in nature and is carried out in close collaboration with public organizations (e.g., housing associations), industry and academia. In doing so, we follow the Socially Responsible Building (SRB) approach, which revolves around a human-centered and integrated approach. We use design thinking and participatory design methods as a basis for this. Validation of our research projects takes place in Living Labs and Field labs, where during the innovation cycles, we slowly scale up from an experimental test environment (for example, in a specially designed research house in Arnhem, Netherlands) to a real-life setting (for example, at a housing association or healthcare organization).

The Dutch Royal Institute of Engineers (KIVI) considers this holistic approach to be of great (social) relevance – one of the reasons for awarding our research group the KIVI-Chair in 2016. KIVI values our comprehensive approach, which is reflected in the expertise present in the research team in areas such as architecture, building technology, sociology, health sciences, industrial design, planning, ethics, (electrical) engineering, and (circular) design.

The Chair Architecture in Health operates at the intersection of the building industry and healthcare, two industries subject to significant transformations. The urgency of transition is strongly felt within both sectors, leading to many

210

initiatives to approach research from the day-to-day practice. We are closely involved in research initiatives in various (construction) companies and (healthcare) organizations and support these organizations with field research as well as advice and monitoring. The chair members are also actively involved in embedding and anchoring (acquired insights from) research in education.

Figure 2
With sensory stimuli, such as the smell of bread spreading in the kitchen while projecting a sandwich on the wall, the smart house in Arnhem contributes to the self-reliance of seniors with dementia.

The Empathic Home: A hybrid learning environment

The 'Empathic Home' project is an empirically grounded illustration of our research. In collaboration with over thirty companies and organizations, this research facility was built in the city of Arnhem (www.empathischewoning.nl). The project started in 2014 and has since grown into a hybrid learning environment aimed at the next generation of smart homes. This home, which has recently attracted much attention from media, serves as a laboratory for researchers and students, a showcase for practice-based research, and a source of inspiration for healthcare, construction, and engineering professionals.

In this experimental home, we study how technology can support both seniors (with dementia) and their informal carers. In this 'empathic' home, human-centered design solutions are designed to take over some of the tasks of the informal caregivers by encouraging older people with dementia to remain active for longer. Sensors, projections, light, and sound signals combined with the architecture of the building guide independently living seniors with dementia in their rhythm. In an iterative design process, the

211

Figure 3 (left)
Dinner time. The arrow on the floor lights up, and the directions in the kitchen help the inhabitant prepare the food.

Figure 4 (right)
The students of the studio Smart Healthy Environment use research through a design approach to study which (new) housing topologies can contribute to social cohesion in an urban or rural context. (Students: L.Y.S. Damen, A.N. Duman, B.K. Tunç, and D. van der Velde).

'Guiding Environment' prototype for this interactive home is designed, tested, and improved until a second prototype can be implemented. The developed products are tested by older people (with dementia). In this, collaboration is being sought with various (care) organizations. The early prototypes of the developed products have been installed in the Empathic Home; the improved prototypes are being tested in some nursing homes. Students from the HAN University of Applied Sciences, Amsterdam University of Applied Sciences, and the Technical University of Eindhoven are involved in this research. In conducting design research, they contribute to the socio-technological solutions for an empathic living environment for this target group.

An example of design research carried out in our Empathic Home project is the 'Edible Wall': a green wall with edible plants that you can adjust to the right height to allow people who have difficulty getting out of the house to garden in their living room. We aim to increase living comfort and keep older people (with dementia) moving and active. The 'COOK3R' has also been developed for this pilot home: a cooking aid that helps the senior prepare meals independently. This appliance shows the proper order of pots and pans to be used when cooking.

Living communities in shrinking regions

The living facilities in Dutch villages are decreasing, and the identity and attractiveness of villages have come under pressure lately. At the same time, since the introduction of civil society in the Netherlands, a growing number of people have participated in local policymaking and organize themselves into civic bodies to improve their living environment. Subjects such as housing and care, which are closely linked to the viability of a community, often form the core of these citizens' initiatives. The project 'Living Communities in Viable Neighborhoods' tackles this problem in a multidisciplinary way, with project partners on both sides of the Dutch-German border in four Dutch and four German villages. In this project, our research group collaborated with the Hochschule Rhein-Waal and local communities in the field of livability (www. euregio.org/action/projects/item/103/krake/).

The project focuses on how the spatial living environment of small communities can be designed to boost the quality of life in the village. The aim is to develop design guidelines for the authorities and designers to support the planning and designing of future-proof rural environments. The gained insights are also translated into publications in both academic and professional journals. In addition, students have been introduced to this theme in various design studios, where they designed solutions for enhancing the residential livability in these villages.

Future developments

Architecture in Health's research projects are usually commissioned by public parties, such as housing associations and care organizations, and are carried out in real-life settings (Field labs), where the multiple stakeholders with different backgrounds come together. In the meantime, our research group and the Eindhoven University have been running eleven Living Labs throughout the Netherlands. These Living Labs are part of the learning community Dutch Empathic Environment Living labs (DEEL Academy), a knowledge platform that develops and shares (DEEL means 'to share' in Dutch) knowledge in the field of 'The New Living.'

213

Figure 5
A collage of the ongoing field lab projects that are brought together in the learning community Dutch Empathic Environment Living labs (DEEL).

An example of the research we conducted in DEEL Academy is the project 'Room for Encounter', commissioned by housing corporation TALIS. The project examines the effects of shared meeting spaces in social housing on stimulating social interaction among (older) residents.

Also, the research group is responsible for several PhD research projects. Some examples are: we, together with the care organization Siza and Academy Het Dorp, are researching (technological) means that increase the self-reliance of their residents with non-congenital brain damage and respiratory support. Another example is the PhD study 'Shared Spaces, Shared lives' in collaboration with the housing association Woonzorg Nederland that explores socio-spatial (design) guidelines and strategies for communal living for seniors in the social housing sector.

The know-how developed by our research group is being widely shared, mainly through our annual conference 'The New Living' jointly organized with de TU Eindhoven, policymakers, managers of care organizations, and housing corporations. Our lecturers and researchers also share their knowledge with students from several HAN departments, such as Built Environment and Engineering. By doing so, we aim to increase our students' awareness of the broader societal context in which they will later work and their ability to research to gather and evaluate relevant, existing knowledge.

Direct knowledge transfer from our research projects to education is carried out in Smart Healthy Environments (SHE) studio. In addition, lecturers from other departments and fields regularly contribute to our educational activities. The aim is to ensure that students' assignments in SHE are better aligned with the Living Lab research projects. This gives students a better understanding of the real-life projects and front-row seats to how research is conducted in practice.

Masi Mohammadi

HAN University of Applied Sciences

Prof.dr.ir. Masi Mohammadi is the Leading Professor in the Academy of the Built Environment at the HAN University of Applied Sciences, where she heads the research group Architecture in Health. Also, she is a Full Professor of Smart Architectural Technologies at the Eindhoven University of Technology. As the principal investigator and leader of the 'Dutch Empathic Environment Living labs' – a nationwide collaborative community consisting of industry, housing, and care organizations – she aims to pilot and empirically evaluate smart homes and neighborhoods. Furthermore, she has served as chair or board member of various (inter)national committees and research networks, as a board member of a European committee on 'Active Aging & Design,' as a member of the Board Science, Technology and Society of The Dutch Royal Institution of Engineers, and as visiting Professor at the University of Technology Sydney.

Seducing the conshuman

Design methods as fertile breeding ground for the food and agriculture industry

Antien Zuidberg

The rapid population growth after WWII demanded an increase in food quality and more efficient food production. To support this, technological-oriented study programs such as Food Technology were created. From the 1980s, designers have also started to design food in their own typical way.[1]

A new higher professional education course, Food Innovation, was created in 2004, to address the gap between food technology and marketing. In addition, the course also focuses on creativity and design subjects to develop food concepts, creatively responding to the needs of *conshumans* (people who consume food). Unlike functional product features such as nutritional value and safety, these food concepts give new meaning and value to food (design-driven innovation).[2]

Fifteen years later, it is time to assess the state of play in applied design research in the food and agricultural industry. Where do we come from, where are we now, and what are the challenges for the future?

1. Marielle Borderwijk and Hendrik Schifferstein, "The Specifics of Food Design: Insights From Professional Design Practice," *International Journal of Food Design* 4, no. 2 (1 August 2020): 101–138.

2. Roberto Verganti, *Design-Driven Innovation, Changing the Rules of Competition by Radically Innovating What Things Mean* (Boston, MA: Harvard Business Press Books, 2009).

3. https://www.hashoge-school.nl/hbo-opleidingen/food-innovation-den-bosch, consulted on 3-5-2021.

4. Hasso Plattner Institute of Design, *An Introduction to Design Thinking: Process Guide* (Stanford: 2010).

5. Tim Brown, "Design Thinking," *Harvard Business Review* 86, no. 6 (June 2008): 85–92.

6. Herbert Simon, *The Sciences of the Artificial, Third Edition* (Cambridge, MA: MIT Press, 1996): 111.

7. Jeanne Liedtka, "Perspective: Linking Design Thinking with Innovation Outcomes Through Cognitive Bias Reduction," *Journal of Product Innovation Management 32, no. 6 (25 March 2014)*: 925–938, https://doi.org/10.1111/jpim.12163.

8. Guido Stompff, *Design Thinking- Radicaal Veranderen in Kleine Stappen* (Amsterdam: Boom uitgevers, 2018)

9. Kees Dorst, Frame Innovation: Create New Thinking by Design (Cambridg: MIT Press, 2015).

Where are we now?

The Food Innovation course is now over 15 years old and has become a well-established name in the Dutch food industry. Food Innovation alumni are active as food professionals in existing and new companies; they play an increasingly important role in the food industry in the area of food innovation. The program provides three main specializations in the field of design, packaging, and marketing.[3]

The design methodology that is taught, is a melting pot of marketing, product development, and graphic design. Years 1 and 2 of the program are based on the backbone of the *Food Innovation model* (see Figure 1), which contains many elements of design thinking.

- To a large degree, the phasing follows the Stanford Institute for Design's design thinking model.[4]
- The program uses Tim Brown's definition 'design thinking is a human-centered approach to innovation that draws from the designers' toolkit to integrate the needs of people, the possibilities of technology and the requirements for business success', with terms such as feasible, viable and desirable.[5]
- The model is aimed at changing the current situation into a new desired situation, according to Herbert Simon.[6]
- The model is aimed at users (conshumans).
- The model uses market elements in the *discover* and *develop* phases, such as carrying out an internal and external analysis and using a SWOT table to achieve strategic *innovation opportunities*, as well as looking at commercial feasibility and the business model in the develop phase.
- The model has an additional *deliver* phase to actually bring concepts to market, which fits in well with the design thinking framework of Darden Business School 'what works'. [7]

The Food Innovation Model fits seamlessly with the 'research *for* design' and 'research *by* design' typing that Guido Stompff mentions as research to create the best design and research to validate the design.[8] Important aspects of design thinking such as iteration or continuous reflection on the design, and the framing of possible solutions (Dorst [9] and Stompff [8]) have been somewhat neglected in the Food Innovation Model and could make the model even stronger.

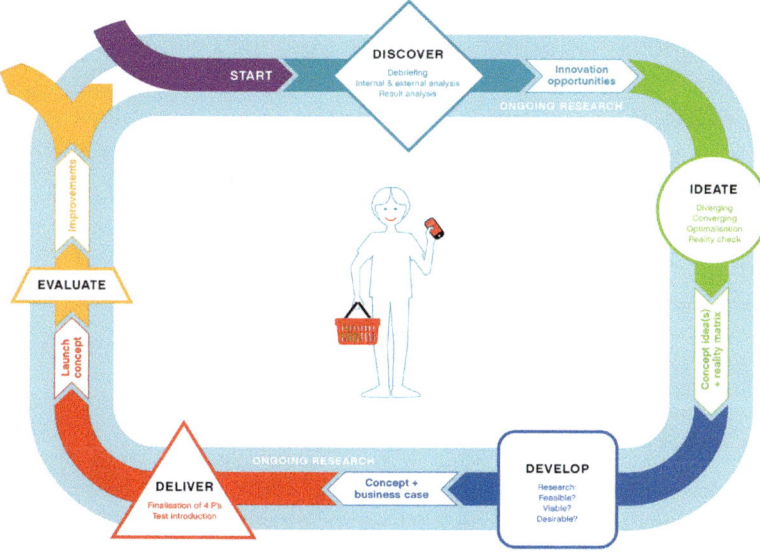

Figure 1
Food Innovation Model. The process of concept development, © HAS University of Applied Sciences, 2016.

Developments in the food industry: Trends & transitions

The developments in the food industry in recent years have focused on convenience, health, and sustainability. We see an increasing role for the transitions that are seen world-wide as fundamental challenges for humans to improve the world, e.g., the United Nations Sustainable Development Goals. In the Netherlands, conshumans are used to having a wide range of affordable, tasty, and convenient food products. The big challenge is to seduce these same conshumans to start eating healthy and sustainable foods.

Over the last twenty years, the food industry has invested largely in technological developments such as sugar reduction, salt reduction, and meat substitutes. By now, there is a wide range of, for example, acceptable plant-based meat substitutes in our supermarkets, and these products have seen a substantial growth in sales.[10] However, according to Wageningen University & Research (WUR) research in 2019, the Dutch population still consumed as much or even more meat than before.[11]

The situation is even worse for products based on alternative, sustainable protein sources such as insects and algae. The production has increased, several products have been marketed, and legislation is being amended. People are curious, but do not yet embrace products with alternative protein sources. Meanwhile, hiding the 'less attractive (read tasty) ingredients' (insect meal in bread, vegetables in sausage rolls) is slowly becoming the norm, to entice the conshuman to make a small step. The thought behind that is that, once people are used to this, they can then take another step. However, this way, we will not achieve the Sustainable Development goals by 2030.

Food transitions' wicked problems

There are several complex human, systemic and wicked problems underlying the desired food transitions. Design thinking, but also systems thinking,[12] because of the holistic and creative approach, have greater chances of succeeding in solving these 'wicked problems' than the classical ways of thinking. What are those 'wicked problems' precisely? Below are some examples of potential foods that could be made more sustainable.

1) We have a different relationship with our food than with our car

A new car or solar panels on our roof is something different from food: we put food in our mouths, in our body, it is essential for our health, and we have a much more intimate relationship with it than with products such as cars. We are used to blindly trusting the food in the supermarket because we have devised systems to ensure its quality. However, we continue to persistently mistrust certain types of food, such as algae and insects. Moreover, we see the mistrust of the major food producers growing. To secure our trust, there is a great need for transparency about our food and where it comes from.

2) We derive our (cultural) identity from eating food

What we eat is determined strongly by our culture and our upbringing. People derive their identity from rituals that include food, and what we don't know, we won't eat. After all: 'What is unknown is unloved.' A good example is the consumption of dairy and meat in the Netherlands. For years, we have heard that dairy and meat are the best food sources, and the idea is that eating meat is linked to manliness. This is reinforced by advertising where meat-eating is portrayed as 'manly'. To give the protein transition a greater chance of success, food, and eating meat in particular, must be separated from our traditional identity values.

3) The temptation of the price and liberal thinking

Many problems can easily be solved by playing with pricing: a 10% price reduction works as well as the best health nudge.[13] Less healthy products such as pizza, fries, burgers, and sausage rolls are more appealing than healthier fruit and vegetable products; price plays a significant role. In general, sustainable food is also more expensive. However, reducing the price of fruit and vegetables, or making meat products more expensive, runs counter to our Dutch liberal principles not to intervene in market forces. Politics may come to rethink this when examples from other countries prove to work well, such as the sugar tax in the United Kingdom.

4) We do not (yet) have a fair price system

The current Dutch food system is out of balance after years of subsidizing and price wars between supermarkets. The competition is fierce and to get that low price, some of the links in our food system chain have been struggling. Farmers in particular do not get a fair price, which prevents innovation for the sake of sustainability. Moreover, actual environmental costs are not yet included in food products. In animal products in particular, a fair price could accelerate the protein transition.

10. Pepijn de Lange, "Nederlanders Eten Van Alle Europeanen de Meeste Vleesvervangers," *De Volkskrant* (10 May 2021).

11. Hans Dagevos, David Verhoog, Peter van Horne and Robert Hoste, *Vleesconsumptie per hoofd van de bevolking in Nederland, 2005–2019, Nota 2020–078* (Wageningen Economic Research, September 2020).

12. Anu Manickham and Karel van Berkel, *Wicked World: Complex Challenges and Systems Innovations* (Groningen: Noordhoff Uitgevers, 2020).

13. Romain Cadario and Pierre Chandon, "Which Healthy Eating Nudges Work Best? A Meta-Analysis of Field Experiments," *Marketing Science* 39, no. 3 (May June 2020): 459 – 665, https://doi.org/10.1287/mksc.2018.1128.

14. Pieter Jan Stappers and Elisa Giaccardi, "Research Through Design," in *The Encyclopedia of Human-Computer Interaction, 2nd edition, eds.* Mads Soegaard and Rikke Friis-Dam (Aarhus, Denmark: 2017): 1–94.

15. Verheijen, L., Praasterink, P. Giezen, P, van Aken, S., and Riedesel, A., *Student Manual: A Food Systems Approach: A Toolkit to Unravel Complexity* (Research group Future Food Systems, HAS University of Applied sciences, 2020).

Shift in applied design research in food & agriculture

Applied design research in food & agriculture is research aimed at creating the best design (concepts, services and systems) for conshumans. To solve the challenging food transitions and solve the 'wicked problems', some of which have been outlined above, we need a strategy shift: from research *for* design to research *through* design.[14] The research is increasingly focused on designing interventions to learn how people (both conshumans and employees in the food industry) are dealing with the new desired situation and how they can be seduced to behave in more sustainable and healthier ways. There is more need for knowledge about human behavior, seduction and nudging of people and companies towards the necessary food transitions. Linked to this, we will be performing more systemic research, focusing not only on a concept or event, but also on underlying systems, such as organization, structures and people's view of the situation.[15]

An example of applied design research in food & agriculture

One of the tools that the Design Methods research group has worked on in the field of food over the past three years is an inspirational seduction model for meaningful innovation in food & agriculture. The seduction model (see Figure 2) has been developed from the idea that healthy and sustainable food concepts need more seduction to influence conshumans towards healthy and sustainable behavior. The aspects explained in the model are based on behavioral literature, marketing, food-design experience, and practical case studies. It is an excellent example of research through design.

This model aims to provide inspirational ways to design meaningful food & agriculture innovations, making them as seductive as possible from the point of view of the conshuman. The model highlights four aspects that affect seduction. In the model, these four aspects are portrayed by an analogy to how bees are attracted by flowers. The four aspects are:

1. the bees are the conshumans who are being seduced;
2. the flower is the sustainable or healthy concept;
3. the field in which they grow is the place of purchase, with multiple flowers (alternative solutions);
4. the magic of the experience, the communication, and other tools to seduce conshumans (in an ethical way) to change their behavior.

Seduction Model

1. the **bee** = the conshuman
2. the **flower** = the food concept
3. the **meadow** = the place of purchase

4. the **magic** = experience, communication and influencing behaviour

Figure 2
The inspirational temptation model for valuable innovation in food & agriculture.

In a previous version, the seduction model has been tested for three food cases with HAS students. The first case was the development of packaging for a commercial insect burger. The students researched the most promising target group and the best communication on the packaging and website to get the target group interested. The second case was a re-design of the HAS cafeteria, run by Appél Catering. The researchers looked into what makes a healthy cafeteria look as appealing as possible for the broad target group of HAS students. In the third case, the seduction model was used to create a banana with a 'true price' (fair price in which

16. Joan Ernst van Aken and Daan Andriessen, eds., *Handboek Ontwerpgericht Wetenschappelijk Onderzoek; Wetenschap met Effect* (The Hague: Boom Lemma Uitgevers, 2011): 151–152.

17. Tonnie van der Zouwen, *Actieonderzoek Doen: Een Routewijzer voor Studenten en Professionals* (Amsterdam: Boom Uitgevers, 2018).

18. Antien Zuidberg, *What U Design = How U Design*, Inaugurele Speech ('s Hertogenbosch: HAS University of Applied Sciences, 2020).

all hidden costs are taken into account) as appealing as possible to the conshumans. In all three cases, designs were created that will be tested as soon as the Covid pandemic permits it. Students and lecturers have provided feedback on the seduction model and how it can be used: a final design and manual to implement it in HAS courses is now being developed.

The future of applied design research in food & agriculture

Although the Food Innovation program and the Food Innovation model have proven themselves to the food industry outside the HAS, the traditional food & agriculture research world is still skeptical about design thinking methods. They want to know what it means exactly and what it offers in addition to traditional research methods. To what extent is applied design research actually research? It is not yet sufficiently recognized that the combination of technology and design research leads to the best design solutions for the future food and agriculture practice.

We hope that these questions will be answered when we solve the 'wicked problems' of food transitions, which cannot be solved only by technology. The demand for proof that applied design research and design methods are effective will gradually become less relevant. When we develop solutions for food issues using applied design research, we are dealing with pragmatic validity: **16** 'the extent to which the research results lead to actions that will produce the desired effects in the future'. It is also expected that conshumans, individuals, and society will be more involved in developing solutions for food transition issues, also known as co-creation, participatory, or action research.**17**

In the coming years, the Design Methods in Food research group will focus on design through research: designing interventions for healthy and sustainable concepts, using the seduction model and themes such as sustainable packaging and preventing food waste.

Antien Zuidberg

HAS University of Applied Sciences

Dr.ir. Antien Zuidberg studied food technology and has a
PhD from Wageningen University & Research. For eleven
years, she worked at the dairy company Campina, where she
worked on developing and applying proteins in food products, among other themes. In 2004, she became a professor
at HAS University of Applied Sciences, for the Food
Innovation program. After seven years as program coordinator for Food Innovation, she became a Professor of Design
Methods in Food in 2019. She is convinced that concepts
that contribute to healthy and sustainable food transitions
can be marketed more successfully using design methods.[18]

225

PART 5:
THE
TASK FOR
APPLIED
DESIGN
RESEARCH

"An explorer can never know
what he is exploring until
it has been explored"

~ **Gregory Bateson**

Something old, something new

How the pillars of design changed dramatically in the last forty years, but the challenges remained the same.

Karin van Beurden

'We must recognize the obvious. It costs more to produce our present forms of ugliness than to create better alternatives. We will be forced (like it or not) towards better, saner and more energy-saving tools and devices, simply because we cannot afford any other kind.' ❶

1. Victor Papanek and James Hennessey, *How Things Don't Work* (New York: Pantheon Books, 1977).

Visiting Kansas City in 1979, I enrolled in a design class with Victor Papanek, who was the head of the Design Department of the Kansas City Art Institute at the time. The six weeks before I had done an internship at the design department of Plantronics in Santa Cruz as part of the Industrial Design course at TU Delft. At Plantronics, I designed headsets intended for mass or serial production. Two different experiences that, in my opinion, seamlessly interlinked.

Industrial Design Research Group

Since 2004, I have been professor of Product Design at the Industrial Design research group, part of Saxion Academy Life Science, Engineering & Design. As a University of Applied Science professor of the first hour, I have shaped this position, and my traits as a designer proved to be an advantage. For example, taking stakeholders as a starting point, but also being fond of changes and being able to deal with uncertainties. My industry-developed standard question 'Yes, but how can it be solved?' also proved to be the motto within the context of the university of applied sciences.

The pitch of the Industrial Design research group is: *'The current social challenges require creative solutions that meet human needs. At the same time, many new technologies are being made available. How do we design meaningful and socially relevant products with these new technologies?'* We answer this question through three research lines.

Research line 1 – *People in Design* is aimed at designing products (and services) in such a way that they align with the real needs and experience of users. Target groups with which experience has been gained are diverse, ranging from caregivers to firefighters and women after a mastectomy.

Research line 2 – *Technology in Design* aims to make knowledge about innovative materials and technologies accessible and translate this into innovative applications. In 2014, we won the SIA [2] Award for the best practice-oriented research with the 'Materials in Design' project.

Research line 3 – *Sustainable in Design* deals with research on how product development and innovation can contribute to a sustainable society. This research line focuses on minimization (longer lifetime and reduction of material and energy consumption), re-use, and recycling.

Design in Motion

In Victor Papanek's view, in the late 70s industrial designers had produced many expensive but awkward and ugly products.[3] And that while at that time, industrial designers followed the Bauhaus ideal to make beautiful and practical design accessible to everyone. The Industrial Design course in Delft was also based on the Bauhaus objectives:[4] creating high-quality technical and esthetic products, with an emphasis on functionality.

In line with this, at the start of the research group in 2004, we focused on 'supporting companies in innovative product development of industrial products for consumers or business to business, which are manufactured in series and/or mass production'. However, with people as the starting point, taking Papanek's critical note into account.

Since then, the domain has been very much in motion, which has had a significant impact on the activities of the research group. Nowadays, a product is usually no longer only physical, but also often has a digital component. Due to technological developments such as additive manufacturing, the size of the series is becoming increasingly smaller. This brings us back to the 'craft of the past', with its unique products. Only now are we producing personalized products on an industrial scale. The ownership of the products is also changing. In the old days, you would buy a lamp; now you buy light; the lamp remains the manufacturer's property. Back then, the industry's initial demands were the main driving force for innovation; nowadays, social challenges are often the starting point. [5] [6] Solutions require collaboration across an entire chain, including societal organizations.

What has not changed is the holistic, the integral approach.[7] The ability to translate complex and sometimes contradictory requirements and research results into possible solutions, 1+1=3, is still characteristic. This puts the design process at odds with (scientific) research. After all, the solution is the product of the designer's thinking process. They make some implicit choices: give two designers the same problem and they will each come up with a different solution, where one solution is not necessarily better or worse than the other.

The belief that the designer can play a role in realizing a better world is still there, as the discussions within the Network Applied Design Research show, just like their focus on change. The ability to imagine that future, to look ahead, and deal with uncertainties is still an essential element.

2. The Taskforce for Applied Research SIA is committed to promoting more and even better applied research by universities of applied sciences. It is part of the Netherlands Organisation for Scientific Research (NWO) and is financed by the Ministry of Education, Culture and Science.

3. Victor Papanek, *Design for the Real World; Human Ecology and Social Change* (St Albans : Paladin, 1974).

4. Walter Gropius, *The New Architecture and the Bauhaus* (Cambridge, MA: The MIT Press, 1965).

5. The Dutch Ministry of Economic Affairs and Climate, *Missies voor het topsectoren- en innovatiebeleid* (The Hague, 26 April 2019).

6. Linda Rindertsma ed., *Kennis- en Innovatieagenda voor de creatieve industrie 2020–2023* (Eindhoven: TKI CLICKNL, 2020).

7. Nigel Cross, "Designerly Ways of Knowing: Design Discipline Versus Design Science," *Design issues* 17, no. 3 (2001): 49–55.

231

8. Pieter Jan Stappers and Elisa Giaccardi, "Research Through Design," in *The Encyclopedia of Human-Computer Interaction, 2nd edition, eds.* Mads Soegaard and Rikke Friis-Dam (Aarhus, Denmark: 2017): 1–94.

Design & research

In the research group ID, design and research are always closely interwoven. We distinguish between research *for* design, research *into* design, or research *through* design.[8] In the beginning, the emphasis of the research group was on research *for* design, but now our portfolio contains all three variants.

Research for design is about collecting knowledge that is important for the development of a particular product or service. For example, the research line *People in Design* involves a lot of quantitative research into the wishes and requirements of target groups, e.g., with the help of personas and various forms of user-centered research. On the other hand, the research line *Technology in Design* focuses on determining the functional properties of innovative materials and technologies and their applications. This is done through literature study and experiments, in which scientific findings are explored and translated into design guidelines.

Figure 1 & 2
Proud Breast: an external breast prosthesis integrated in lingerie.

Proud Breast

The current external breast prostheses for women who have undergone a mastectomy are heavy, sweaty, rubber-like items. In collaboration with Proud Breast (a start-up company), researchers looked into how women use and experience these external breast prostheses. Interviews revealed two main reasons for wearing an external breast prosthesis: to experience a personal sense of normality and the avoidance of uncomfortable situations with other people (taboo).

However, a literature study and discussions with taboo experts show there is a third strategy: full acceptance by showing it openly. This last strategy led to discussions with students about the designer's own responsibility versus the results of the user research. None of the women who participated in the study had indicated that 'showing' was an option. The question is whether this is because this option did not suit them, or because there are still no proper solutions that positively confirm the women's self-esteem? The solutions shown during the Dutch Design Week 2018 evoked many emotional discussions with the visitors.

9. Gerard van Os and Karin van Beurden, "Emogram: Help (Student) Design Researchers Understanding User Emotions in Product Design," *Proceedings of the 21st International Conference on Engineering and Product Design Education (E&PDE 2019)* (Glasgow, September 2019), https://doi.org/10.35199/epde2019.44.

Loc2use

Lab-on-a-Chip (LOC) devices are expensive and can only be used under special conditions. However, LOC devices show great potential for use outside the laboratories, such as in healthcare and forensic investigation. The Saxion research groups NanoBio and Industrial Design work together with corporate partners to make LOCs cheaper and suitable for 'everyday' applications.

By creating current and future usage scenarios with users in the healthcare sector and forensic investigators, real user situations are explored, and requirements identified. At the same time, experiments with 'normal' production methods are performed to determine their potential for producing LOCs. By bringing the use cases together with the technological possibilities, new solutions for microfluidic devices have been realized.

Research into design involves research into (the improvement of) the design process or the development of design-related methods. This is often a by-product of the other activities in the research group ID. For example, we try out new scientific methods in practice. By reflecting on this, we acquire knowledge about the applicability of these methods. In some cases, it will lead to variants or further developments.

Emogram

The Emogram [9] was developed out of frustrations that it was not possible to record the primary – emotional – associations of people with a product using standard interview techniques. Without the usual mixture with cognitive reasoning of the answers, the primary responses are ranked by importance.

Selectietool (selection tool)

This is a model designed to come up with innovative applications for new technologies. It is a variant of the Delft Innovation Model, focusing on the properties and strengths of the new material or technology instead of on the company. By linking these strengths to challenges, promising new applications can be defined.

Research through design raises the most questions in the formal research world, but it is very characteristic of design research. (Prototypes of) concepts or designs are developed and then used with the intention of acquiring knowledge. This is where the design and research approaches are most closely intertwined.

In the research group ID, much knowledge is developed on concrete cases. For example, the possibilities of new materials or technologies are discovered by designing concrete applications for them, such as in the research into 3D metal and concrete printing. That goes hand in hand with experimenting: how slanted can you print? And how high? Reflecting on multiple cases generates both generic and specific knowledge. Sometimes it is necessary to design first, to get a clear picture of the requirements. This is the case, for example, with the application of completely new technology, such as in the Drone Robot project.

Figure 3
Experiments with 3D metal printing.

235

Figure 4
Green Dome Case, 3D
concrete printing and parametric design.

Green Dome

Unlike other research groups, the Industrial Design research group and its partners, do not see the challenge in printing an entire house, but in looking for interesting applications where the specific benefits of 3D concrete printing come into their own. The possibilities are explored by means of cases, such as a 'vispassage' (fish pass in a river) and the Green Dome. The Green Dome, a housing for a high-rise green waste collection point, consists of 58 parametrically designed concrete blocks. The computer calculates the ideal shape based on preconditions, such as that the blocks must not be heavier than 50 kilos, so that you can lift them with two people under occupational health and safety regulations.

Drone Robot

*How would a farmer use a **drone**? What are the requirements? Interviews showed that farmers had trouble imagining the potential of this new technology, let alone say something useful about the requirements. However, by allowing the farmer to experience various scenarios of the drone through a fictitious control panel, the discussions yielded valuable results. The researchers were able to formulate an initial list of requirements.*

Figure 5
Green Dome Case, 3D
concrete printing and
parametric design.

10. Papanek, Design for
the Real World.

Research as a condition

Victor Papanek's quote, which started this story, summarizes how I see the designer's role, or if you like, as a director of change. You cannot rely on your own limited perception and knowledge. Design research is a precondition for making well-founded choices for change, ensuring that the right products are developed, but even more important, that the right problems are addressed.

This does not mean that if you base everything on research, you will automatically do the right things. The Proud Breast example illustrates that the designer/researcher, with their critical mind, also has their own responsibility. After all, if you were to leave it to the women interviewed, there would be no need for taboo-breaking, self-esteem-improving solutions.

Viktor Papanek did not think much of designers, but fortunately, he gives the profession a second chance, under strict conditions: 'Design must become an innovative, highly creative, cross-disciplinary tool responsive *to the true need of men*. It must be more *research-oriented*, and we must stop defiling the earth itself with poorly-designed objects and structures.' [10] In this statement, he ascribed a vital role to a research-oriented approach. More than fifty years later, we are still working on this.

11. Michael Rubenstein, Alejandro Cornejo, Radhika Nagpal, "Programmable Self-Assembly in a Thousand-Robot Swarm," *Science* 345, no. 6198 (15 Aug 2014): 795–799, https://doi.org/10.1126/science.1254295.

Time for change

One by one, the pillars supporting the field of industrial design have changed dramatically. What will be the next change? I guess that it will be the integral, holistic approach. At Harvard, the Self-Organizing Systems Research Group is working on insect-like little robots that find their place in a swarm, not because they are sent there according to an all-encompassing design, but through mutual communication and coordination.

To me, self-driving cars and 'a thousand-robot swarm' [11] are metaphors for a radically different way of designing. There are more signs: for instance, the developments in parametric design. What will be the 'definition' of a new developed product? Is it a kind of digital framework that is personalized using scan data? This is a fascinating theme, whose design and design research implications are still entirely unclear.

At the same time, the question remains whether, after forty years, there is anything that has changed. After all, there are still so many poorly designed products, such as all those that have simply disregarded the specific requirements and wishes of women. For example, seat belts and bicycle helmets must have been designed by men, for men. That does not mean that men cannot design for women, but there is still a lot of room for improvement. A great challenge for the new generation of designers and design researchers who will take over from me, is the focus on design and research *'responsive to the true need of man'*. Because it solves something, not just because it is possible.

Karin van Beurden

Saxion Academie Life Scienc

Ir. Karin van Beurden is a Professor of Product Design at the Industrial Design research group, part of Saxion Academy Life Science, Engineering & Design. She has over twenty years of corporate experience in product development and new business creation for consumer and technical products. Since 1999, she has had her own consultancy firm Kompane. She was the director of Kompani BV, a supplier of innovative fire blankets. She has been awarded various GIO awards and patents. Karin also leads Fablab Enschede, an easily accessible digital workshop that is part of the research group. In addition, the research group has several labs, including a concrete and metal printer and a Design Thinking/Usability Test lab. The Kenniskring consists of approximately fifteen researchers, designers, and project leaders. In 2014, Karin won the SIA Award for the 'Materials in Design' project. She is the initiator and, since 2016, chair of NADR, the Network Applied Design Research platform.

239

A letter from the future

An attempt at good ancestry

Jeroen van den Eijnde

To:
Aapeli van den Eijnde
Circular Road
3748 XL Sustainaville

Arnhem, 18 January 2046

Dear Aapeli,

How nice of you to ask me to tell you about applied design research because you are considering training in this field. As your loving grandfather, I will gladly write a letter on this subject, hoping that next year – can you believe it is already 2046 – you can make a well-considered choice.

As you know, around the year 2020, I was intensively involved in the Network Applied Design Research (NADR): a rather curious, quirky but inspiring group of people, who worked on applied design research on a daily basis in their positions as professors at universities of applied sciences. At the time, I was a professor at an art college. This position no longer exists; there are only HESPE professors now.

Back in 2020, higher education was still separated. There were universities, originally intended to train researchers, and universities of applied sciences, preparing students for specifically described professional profiles. This distinction was already outdated at that time. Formulating professional profiles for an art college was a problematic matter anyway.

241

This is not very interesting for you, now that Dutch higher education has been integrated in its entirety into HESPE (Higher Exploring and Scientific Practice-Based Education), including both research and vocational training. However, you might benefit from some historical context, especially to better interpret the term 'applied'. But I will first explain to you something about the concepts *design* and *research*.

About industrial and form-giving designers

I've told you that your granddad trained as a designer at an art academy. Even back then, it was a challenging profession to describe. We were not allowed to call ourselves industrial designers because that was only for those with a university education – indeed, those institutes that at the time trained people to be researchers, not for a profession. There was also the Academy for Industrial Design, but that was for form-giving designers and not industrial engineers. The difference? Long discussions were devoted to this topic, with the industrial designers accusing the form-giving designers of superficial beautification and form-giving designers criticizing the industrial designers for their mostly technical and economical approach. Designs at an art academy were considered a form of applied art. Yes, there is that word 'applied' again. Sometimes I think that if something is not quite what it should be, they put the term applied in front of it: applied research, applied art, and in my time, there was even a school for applied philosophy.

At the Art Academy, almost all design students read the book *Design for the Real World* by American designer Victor Papanek.[1] Papanek introduced his book with a real dig at every design student: 'There are professions more harmful than industrial design, but only a very few of them'.[2] So, in fact, we were quite lucky that we were not allowed to call ourselves industrial designers, but just (product) designers. Pananek even managed to get in a second provocation in his book: 'All men are designers. All that we do, almost all the time, is design, for design is basic to all human activity'. According to Papanek, every planned and organized 'act toward a desired, foreseeable end' – from writing a poem and composing a concert to raising children and baking a cake – is all part of the design process.[3] This also applies to this letter: my aim is to create a text that teaches you something about applied design research. But I would say that when something, design for instance, represents all human, targeted actions, it basically means nothing. Everything always disappears in the nothingness of the unseen, the unseparated. It is everything and therefore not recognized and acknowledged. Moreover, this broad definition may explain the jumble of eccentric members of the NADR network: it included researchers on food, health, biomaterials, innovation networks and co-creation.

Tacit knowledge

Papanek forced me to think (a form of real or applied philosophy?) about what design meant to me at that moment. Based on what I learned at the Art Academy, I concluded that design primarily represents a form of knowledge that you cannot express in language. It is knowledge that you use and acquire by using your body and all your senses: the taste of porcelain, the smell of rosewood, the touch of wool, the sound of a knob turning, the visual perception of something you can use. I learned only much later that this is called tacit knowledge: implicit knowledge that is not text-based, such as intuition and physical routines. Just explain how you learned to ride a bicycle. Someone can show you how, someone can explain it, but in the end, you learned it through falling and getting up: *learning by doing*. A design process is characterized by *trial and error*, two steps forward, one step back. That is what we call iteratively. It's not fast, *but you learn a lot from it.*

1. Victor Papanek, *Design for the Real World. Human Ecology and Social Change* (St. Albans: Paladin, 1974).

2. Papanek, *Design for the Real World*, ix.

3. Papanek, *Design for the Real World*, 3–4.

243

4. Ludwig Wittgenstein, *Tractatus Logico-Philosophicus* (London: Kegan Paul, Trench, Trubner & Co, 1922).

5. Michael Polanyi, *The Tacit Dimension* (New York: Doubleday Anchor, 1966): 4.

Language has always been our main knowledge base for research. Illustrating this is the fact that I am writing you a letter to transfer my knowledge about applied design research to you. However, language is available in different variants. Not every language is suitable for transferring knowledge. This is mainly due to the type of knowledge and the degree to which the language used can be interpreted differently by the reader. That is why the philosopher Ludwig Wittgenstein claimed that 'whereof one cannot speak, thereof one must be silent'. At school, you know that mathematics and physics use a completely different language than, for example, history or economics. Mathematics is seen as the ultimate formal language, in which subjective interpretations are not possible. With his *Tractatus Logico-Philosophicus*, Wittgenstein tried to write a book that can be unambiguously interpreted as a mathematical formula.[4]

Despite my college-prep diploma (with Mathematics and Algebra), completed higher professional training and academic courses, and a doctoral degree, I must confess that Wittgenstein's booklet is still incomprehensible to me. Wittgenstein himself concluded that although his formalized language could transfer certain knowledge objectively, there is also a lot that cannot be said with it. Or, as scientist Michael Polanyi, a contemporary of Wittgenstein, formulated it in 1966: 'I shall reconsider human knowledge by starting from the fact that we know more than we can tell'.[5] He called this the 'tacit dimension', in which unconscious knowledge, based on tradition, inherited practices, implicit values and prejudice or judgments, constitutes a crucial part of scientific knowledge that is considered objective.

Within academic and political circles, sometimes the distinction is made between 'knowledge' and 'applied knowledge' (ergo no real knowledge?), with knowledge representing a scientific, objective 'truth' that can or cannot be applied successfully to practical environments within society. However, Polanyi states that this true knowledge – or perhaps it is better to speak of 'useful' knowledge, as do the pragmatists who pursue it – cannot be acquired without interference from unconscious knowledge from the practical environment. In short, if we do not want to merely explain the world, but above all understand it well, we need not only

Figure 2
NEFFA (Aniela Hoitink) and with Karin Vlug, MycoTEX© jacket, designed as part of the ArtEZ Future Makers project The Future of Living Materials, 2018 (photo: NEFFA).

the academic knowledge that is established in the scientific lab, but also the experiential knowledge that is part of the complex, everyday practice and that, in my time, was mainly researched through so-called *living labs*.

Research the future

But how do you research knowledge that cannot be expressed in words within such a living lab? I think that is best done by emulating the everyday complexity as closely as possible, not with language, but in the form of artifacts that people can experience through seeing, feeling, smelling, tasting. This is an aesthetic experience in the original sense of the word: the doctrine of sensory perception. It is mainly the designer who is trained in the sensory qualities of artifacts. The designer does not use language but prototypes that give a sneak peek into the desired future because design is *'any act towards a desired end'*, as Papanek rightly pointed out.

A prototype is nothing more or less than an aesthetic manifestation of this future. They are knowledge carriers who act as stepping stones towards a desired future situation. That makes applied design research such an interesting domain: you're not researching what already exists but how something better could exist. Most scientists reflect on existing phenomena; applied design researchers investigate potential futures by making them tangible by using prototypes. It is a kind of science fiction, but with design as the most essential instrument. So, design fiction, but with the intention of making the fiction a reality.

But what future do we want? In my time, we started realizing that the earth was in crisis due to two phenomena that the human mind can hardly comprehend: long-term thinking and exponential growth. To give you an example of the latter: if the number of drops of rain above the Ajax soccer stadium (which you know very well) would increase exponentially every second (i.e., one drop in one second, two drops in two seconds, four drops in three seconds, eight drops in four seconds, etc.), how long would it take for that whole Arena to be filled with rainwater? It would only take about one hour!

With regards to the first phenomenon: can we determine, for the next fifty years, the consequences for man and the environment of our everyday activities, such as frequently eating meat, buying new clothes every season and regularly booking an air trip for work and holiday? Since about 1850, the world has experienced exponential growth in prosperity. What we were unable to see, for over a century, is that this also has a substantial negative impact on our planet. The climate crisis proved not to be a conspiracy theory devised by a bunch of weird scientists and environmentalists. Instead, it caused an irreversible change that became physically tangible, directly or indirectly. For example, you have never been skating on natural ice, something that granddad frequently did in winter in the previous century.

Figure 3
Frank Kolkman, the interior of the 'Objects for the Sharing Economy' model, designed as part of the ArtEZ Future Makers project Designing for Precarious citizens. Building upon the Bauhaus Legacy, 2020. (photo: Juuke Schoorl)

Research into a sustainable heritage

6. Roman Krznaric, *The Good Ancestor. How to Think Long Term in a Short-Term World* (London: Penguin Group, 2020).

When I was a lecturer, the Netherlands aimed to have a completely circular and climate-neutral economy by 2050. It is 2046 now, and you see how far we have come. At the time, it was my ambition to be a part of this effort. A philosopher from that era thought the most crucial question you had to ask yourself was 'are we good ancestors?' [6] Or, in other words, are we now acting in a way that guarantees a good future for our grandchildren (you, in this case)? I leave the judgment to you, but you should know I did my best.

Back to applied design research. Based on what I have told you, I consider it to be research into a desirable future using design in the form of prototypes, which are applied as best as possible to everyday practices. As a professor in Tactical Design at ArtEZ University of the Arts, I have tried to shape it as well as possible. NADR's strength was that we were close to these everyday practices through our well-organized networks of companies, societal organizations, and authorities/governments. Changing from a complex practice to the desired situation requires good collaboration between scientists (for the necessary fundamental knowledge), policy makers (for the decisions to be taken), and designers (for the creation of the required solutions).

I will give you two concrete examples of applied design research that we worked on and show you some prototypes that contributed to some significant changes. One study focused on a new kind of textile made from fungus mycelium: the spores of a mushroom. At the time, the fashion and textile industry was one of the most destructive systems for man and the environment. Back then, it was expected that by 2050, if nothing changed, this system would be responsible for 25% of our CO_2 emissions. Our newly developed textile could grow locally at a large scale, based on organic waste streams, could be seamlessly formed three-dimensionally, had no added toxic substances, and saved huge amounts of water in the production process (the production of one cotton T-shirt costs 2500 liters of water, compared to 12 liters of water for a mycelium T-shirt). After use, the material was entirely naturally degradable and did not create waste (Figures 1 and 2). In my

Figure 4
Frank Kolkman, photo
impression of the outer
facade of the 'Objects for
the Sharing Economy'
model, designed as part of
the ArtEZ Future Makers
project Designing for
Precarious citizens. Building
upon the Bauhaus Legacy,
2020 (photo: Juuke Schoorl).

Figure 4
Frank Kolkman, photo impression of the outer facade of the 'Objects for the Sharing Economy' model, designed as part of the ArtEZ Future Makers project Designing for Precarious citizens. Building upon the Bauhaus Legacy, 2020 (photo: Juuke Schoorl).

time, we would generate 17.5 kilograms of clothing waste per person per year worldwide. Just imagine how many garments that represents on average! Think of the impact our material could have on this system. Unfortunately, the fashion system has not yet started using this material on a large scale. But fortunately, we have already made a big difference in singular use work clothing and hotel textiles. I think that had a significant impact.

Another study focused on vulnerable people in society, partly due to the vastly increased technological possibilities. The so-called platform or sharing economy led to companies such as Airbnb and Uber that, using advanced technology, digitally offered their services, which were then carried out by private parties. This included renting out rooms or offering taxi rides. The government has since put a stop to this. And rightly so, because these companies did not take responsibility for their 'employees', whom they did not formally employ, but who were entirely dependent on them for their income.

One of the applied design researchers devised a concept in which low-income individuals, using smart technology, could offer services for a fee, such as the use of a coffee machine, refrigerator, microwave, and washing machine. The solution was a new type of building facade, where the resident can use these appliances both inside and outside the house (Figure 3 and 4). This building facade was never realized, but the discussion that led to the plan made the government decide to impose stricter rules on these businesses.

My dear Aapeli, these were two attempts to answer the most crucial question at the time (although it is still very current): How can I as an individual, and how can we as a community be good ancestors? I think that applied design research is the best answer. I am therefore looking forward to you continuing my work by training in this area. I hope this letter will positively contribute to your choice.

Your loving granddad Jeroen

Jeroen van den Eijnde

ArtEZ University of the Arts

Dr. Jeroen van den Eijnde studied product design at the Arnhem Art Academy and art history at Leiden University. He obtained his PhD with a study into the theory and ideology in Dutch form-giving education. Since 2016, he has worked as a Professor of Tactical Design at ArtEZ University of the Arts. Van den Eijnde was co-founder and board member of the Design Platform Arnhem. As a consultant, he worked for the Fonds Beeldende Kunst, Vormgeving en Bouwkunst (the Fine Art, Design and Architecture Foundation, now the Stimuleringsfonds Creatieve Industrie – the Foundation for the Creative Industry) and the Raad voor Cultuur (Culture Council). He is currently a core member of NADR and a member of the program council for CLICKNL, the innovation network of the creative industry's top sector.

IN CONCLUSION

"To be able to ask a question clearly is two-thirds of the way to getting it answered."

~ **John Ruskin**

Epilogue

Peter Joore, Guido Stompff, Jeroen van den Eijnde

Of course, we are not the first to write about design research. The multitude of perspectives in this publication reflects an academic discussion about the how, what and why of design research. These articles cover an overwhelming collection of concepts, ranging from *generative research* [1] to *action design research.*[2] This sometimes culminates into a bit of wordplay, for example, if a distinction is made between *practice-led design research* [3] and *design research through practice.*[4] For outsiders, the differences addressed are far too subtle to be understood. In this book, this discussion manifests itself in, for example, the many reflections on research for/in/through/by design.

It is tempting to conclude this search for the meaning of applied design research with a 'final' definition, but we will not do so. Another definition, another pigeon-hole, does not add to the lively discussion. Instead, we want to contribute by mapping the rapidly developing landscape of applied design research. We embrace and appreciate the pluralism that manifests itself. Or, in the words of Richard Buchanan, "one of the great strengths of design is that we have not settled on a single definition. Fields in which definition is now a settled matter tend to be lethargic, dying, or dead fields, where inquiry no longer provides challenges to what is accepted as truth."[5]

The pluralism that manifests itself in the landscape also shows that (applied) design research is not a variant of some research method, or that it is reserved for UX / service / experience design, etcetera. On the contrary, the richness arises because inspiration is drawn from the many different forms of research, such as anthropological field research or action research. And because the subject of design has become ever broader, with the introduction of new forms such as legal or systemic design. As such, applied design research effectively bridges the world of research with the world of design.

Its bridging capabilities may justify its existence, but at the same time, it creates obligations. Pluralism carries the risk to turn into 'anything goes' relativism, making applied design research an empty shell without purpose and quality standards. This book offers a compelling picture of the state of applied design research at the Dutch universities of applied sciences. It raises questions, important questions for its practitioners, as well as for methodologists who focus on research and/or design; these questions may lead to a new research agenda. In conclusion, we name some of them:

- In design research, knowledge is partly embodied by design artifacts, such as prototypes, which raises questions about the underlying ontology and epistemology. *In design research, what is knowledge, and how do we obtain that knowledge?*
- Applied design research is embedded in local practices, which means that accumulated knowledge is situational, but may also be applicable in other situations, at least to some degree. *How do we distinguish and generalize that knowledge?*
- Applied design research uses various existing methods and often adapts them to the design context. This has created various (derived) methods, but *which quality criteria must these methods meet to be valid and reliable?*
- Applied design research almost always involves several stakeholders with different interests and power relations. *How do you handle inherent power relationships between those involved?*
- Design research is based on different world views, values, and standards of involved actors. At the same time, it is intervening and thus cuts into the practical environment, into people's lives. *What are the underlying values and associated ethical standards for applied design research?*

These questions are not so much about the added value of applied design research, its 'relevance', but more about the scientific 'rigor'. By answering these questions, the quality of the research will improve, and applied design research will reach maturity in many more fields of knowledge than we covered in this book!

1. Elizabeth Sanders and Pieter Jan Stappers, *Convivial Toolbox. Generative Research for the Front End of Design* (Amsterdam: BIS publishers, 2012).

2. Sein, Maung K., Ola Henfridsson, Sandeep Purao, Matti Rossi, and Rikard Lindgren. "Action Design Research." *MIS Quarterly* (2011): 37–56.

3. Maarit Anna Mäkelä and Nithikul Nimkulrat, "Reflection and Documentation in Practice-Led Design Research," in *Proceedings of the 4th Nordic Design Research Conference* (Helsinki, 2011).

4. Ilop Koskinen, John Zimmerman, Thomas Binder, Johan Redstrom, and Stephan Wensveen, *Design Research Through Practice: From the Lab, Field, and Showroom* (Amsterdam: Elsevier, 2011).

5. Richard Buchanan, "Design Research and the New Learning," *Design Issues 17*, no. 4 (Autumn 2001): 8.

LITERATURE

- Danah **Abdulla**, *Modes of Criticism 4; Radical Pedagogy* (Eindhoven: Onomatopee, 2019).
- Bas van **Abel**, Roel Klaassen, Lucas Evers, Peter Troxler, *Open Design Now: Why Design Cannot Remain Exclusive* (Amsterdam: BIS Publishers, 2011).
- Manuela **Aguirre**, Natalia Agudelo, and Jonathan Romm, "Design Facilitation as Emerging Practice: Analyzing How Designers Support Multi-Stakeholder Co-Creation," *She Ji: The Journal of Design, Economics, and Innovation* 3, no. 3 (2017): 198–209.
- Joan Ernst Van **Aken**, and Daan Andriessen, eds., *Handboek Ontwerpgericht Wetenschappelijk Onderzoek; Wetenschap Met Effect* (Den Haag: Boom Lemma Uitgevers, 2011).
- Robert **Anderson**, *European Universities From the Enlightenment to 1914* (Oxford: Oxford University Press, 2004).
- Daan **Andriessen**, *Praktisch Relevant én Methodisch Grondig? Dimensies van Onderzoek in het Hbo*, Openbare Les Hogeschool Utrecht (10 April 2014).
- Paola **Antonelli**, and Jamer Hunt, *Design and Violence* (New York: The Museum of Modern Art, 2015).
- Bruce **Archer**, "Design as a Discipline," *Design Studies* 1, no. 1 (1 July 1979): 17–20. https://doi.org/10.1016/0142–694X(79)90023–1.
- James **Auger**, "Speculative Design: Crafting the Speculation," *Digital Creativity* 24, no. 1 (March 2013): 11–35. https://doi.org/10.1080/14626268.2013.767276.
- Ivana **Bartoletti**, *An Artificial Revolution: On Power, Politics and AI* (London: The Indigo Press, 2020).
- Mortaza S. **Bargh**, and Peter Troxler, "Digital Transformations and Their Design – Renewal of the Socio-Technical Approach," in *Hoger Beroepsonderwijs in 2030. Toekomstverkenningen, and Scenario's vanuit Hogeschool Rotterdam*, eds. Daan Gijsbertse, Arjen van Klink, Kees Machielse, and Jeroen Timmermans (Rotterdam: Hogeschool Rotterdam Uitgeverij, 2020).
- Katja **Battarbee**, Jane Fulton Suri, and Suzanne Gibbs Howard, "Empathy on the Edge: Scaling and Sustaining a Human-Centered Approach in the Evolving Practice of Design," *Harvard Business Review* (January 01, 2015).

- Roy **Bendor**, Aadjan van der Helm, and Tomasz Jaskiewicz, eds., *A Spectrum of Possibilities: A Catalog of Tools for Urban Citizenship in the Not-So-Far Future (*Delft University of Technology, 2018).
- Adrie **Beyen**, *Kathalys: Vision on Sustainable Product Innovation* (Amsterdam: BIS Publishers, 2001).
- Linda **Blaasvaer**, and Birger Sevaldson, "Educational Planning for Systems-Oriented Design: Applying Systemic Relationships to Meta-Mapping of Giga Maps," in *DS 95: Proceedings of the 21st International Conference on Engineering and Product Design Education* (University of Strathclyde, Glasgow, 2019).
- Kirsten **Boehner**, Janet Vertesi, Phoebe Sengers, and Paul Dourish, "How HCI Interprets the Probes," *Proceedings of The SIGCHI Conference on Human Factors in Computing Systems* (2007): 1077–1086.
- Kirsten **Boehner**, William Gaver, and Andy Boucher, "Probes," in Celia Lury en Nina Wakeman eds., *Inventive Methods: The Happening of the Social* (New York: Routledge, 2012): 185–201.
- Ian **Bogost**, *Play Anything* (New York: Basic Books, 2016).
- Richard J. **Boland**, and Fred Collopy eds., *Managing as Designing* (Stanford, CA: Stanford Business Books, 2004).
- Marielle **Borderwijk**, and Hendrik Schifferstein, "The Specifics of Food Design: Insights From Professional Design Practice," International Journal of Food Design 4, no. 2 (1 August 2020): 101–138.
- Cennydd **Bowles**, *Future Ethics* (London: NowNext, 2018).
- Rens **Brankaert**, Gail Kenning, Daniel Welsh, Sarah Foley, James Hodge, David Unbehaun, "Intersections in HCI, Design and Dementia: Inclusivity in Participatory Approaches," in *DIS 2019 Companion – Companion Publication of the 2019 ACM Designing Interactive Systems Conference* (San Diego, CA, June 2019). https://doi.org/10.1145/3301019.3319997.
- Rens **Brankaert**, and Elke Den Ouden, "The Design-Driven Living Lab: A New Approach to Exploring Solutions to Complex Societal Challenges," *Technology Innovation Management Review 7, no. 1* (2017): 44–51.
- James **Bridle**, *New Dark Age* (London: Verso, 2018).
- Tim **Brown**, *Change by Design* (New York: HarperCollins Publishers Inc, 2008).
- Jos **van den Broek**, Isabelle van Elzakker, Timo Maas, Jasper Deuten, *Voorbij Lokaal Enthousiasme. Lessen voor de Opschaling van Living Labs* (Den Haag: Rathenau Instituut, 2020).

- Tim **Brown**, "Design thinking," *Harvard Business Review* 86, no. 6 (June 2008): 85–92.
- Tim **Brown**, and Jocelyn Wyatt, "Design Thinking for Social Innovation," *Development Outreach* 12, no. 1 (2010): 29–43. https://doi.org/10.1596/1020-797X_12_1_29.
- Finn **Brunton**, and Helen Nissenbaum, *Obfuscation. A User's Guide for Privacy and Protest* (Cambridge: The MIT Press, 2012);
- Richard **Buchanan**, "Wicked Problems in Design Thinking," *Design Issues* 8, no. 2 (Spring, 1992): 5–21.
- Richard **Buchanan**, "Design Research and the New Learning," *Design Issues* 17, no. 4 (Autumn 2001): 8.
- Romain **Cadario**, and Pierre Chandon, "Which Healthy Eating Nudges Work Best? A Meta-Analysis of Field Experiments," *Marketing Science* 39, no. 3 (May June 2020): 459–665. https://doi.org/10.1287/mksc.2018.1128.
- Violeta **Clemente**, Katja Tschimmel, and Fátima Pombo (2017) A Future Scenario for a Methodological Approach applied to PhD Design Research. Development of an Analytical Canvas, *The Design Journal* 20 (6 Sept 2017): S792 – S802. https://doi.org/10.1080/14606925.2017.1353025.
- Emma **Cocker**, "Tactics for Not Knowing: Preparing for the Unexpected," in: Elizabeth Fisher, and Rebecca Fortnum, eds., *On Not Knowing: How Artists Think* (London, Black Dog Publishing, 2013).
- Dalila Cisco **Collatto**, Aline Dresch, Daniel Pacheco Lacerda, Ione Ghislene Bentz, "Is Action Design Research Indeed Necessary? Analysis and Synergies Between Action Research and Design Science Research," *Systemic Practice and Action Research*, 31, no. 3 (2018): 239–267. https://doi.org/10.1007/s11213-017-9424-9.
- Anke **Coumans**, and Ingrid Schuffelers, "De Relevantie van Artistiek Onderzoek," *ScienceGuide*, 21 June 2017.
- Anke **Coumans**, "Relational Drawing. De Kunstenaar als Antropoloog," *FORUM+ voor Onderzoek en Kunsten*, vol. 26, no. 1, 2019, pp. 38–47.
- Anke **Coumans**, "Ontwerpen in Het Hier en Nu. De Artistieke Attitude in de Zorg voor Mensen met Dementie," *FORUM+ voor Onderzoek en Kunsten* 27, no. 2 (2020): 3–13.
- Petra H.M. **Cremers**, *Designing Hybrid Learning Configurations; At the Interface between School and Workplace*, PhD Thesis Wageningen University (10 February 2016).

- Nigel **Cross**, "Designerly Ways of Knowing," *Design Studies* 3, no. 4 (1982): 221–27. https://doi.org/10.1016/0142–694X(82)90040–0.
- Nigel **Cross**, "Designerly Ways of Knowing: Design Discipline Versus Design Science," *Design Issues* 17, no. 3 (2001): 49–55.
- Nigel **Cross**, "From a Design Science to a Design Discipline: Understanding Designerly Ways of Knowing and Thinking," In *Design Research Now: Essays and Selected Projects*, ed. Ralf Michel (Basel: Birkhäuser, 2007): 41–54. https://doi.org/10.1007/978–3-7643–8472–2_3.
- Marcel **Crul**, Plastic-Free Tourism and Hospitality on Dutch Wadden Islands: Multi-level Design Approaches and Experiences. *Proceedings of European Roundtable for Sustainable Consumption and Production* (Graz, 2021).
- Robert **Curedale**, *Design Thinking: Process and Methods* (Topanga: Design Community College, 2016).
- Hans **Dagevos**, David Verhoog, Peter van Horne, and Robert Hoste, *Vleesconsumptie per hoofd van de bevolking in Nederland, 2005–2019, Nota 2020–078* (Wageningen Economic Research, September 2020).
- Peter **Dalsgaard**, "Pragmatism and Design Thinking," *International Journal of Design* 8, no. 1 (2014).
- Jane **Darke**, "The Primary Generator and the Design Process," *Design Studies 1, no.* 1 (1979): 36–44.
- **Design Council**, *Beyond Net Zero – A Systematic Design Approach* (London: **Design Council**, April 2021).
- Aranka **Dijkstra**, Sybrith Tiekstra, Gertjan de Werk, Peter Joore, "Festivals as Living Labs for Sustainable Innovation: Experiences from the Interdisciplinary Innovation Programme DORP," *Proceedings of European Roundtable for Sustainable Consumption and Production (*Barcelona, 2019).
- Aranka **Dijkstra**, and Marije Boonstra, *Festival Experimentation Guide,* (Leeuwarden, NHL Stenden Publishers, 2021).
- Carl **Disalvo**, *Adversarial Design* (Cambridge: The MIT Press, 2012).
- Brian **Dixon**, *Dewey and Design: A Pragmatist Perspective for Design Research* (London: Springer Nature, 2020).
- Kees **Dorst**, "Design Research: A Revolution-Waiting-To-Happen," *Design Studies* 29, no. 1 (2008): 4–11.
- Kees **Dorst**, "The Core of 'Design Thinking' and Its Application," *Design Studies* 32, no. 6 (2011): 521–532. https://doi.org/10.1016/j.destud.2011.07.006.

- Kees **Dorst**, *Frame Innovation: Create New Thinking by Design* (Cambridge: MIT Press, 2015).
- Anthony **Dunne**, and Fiona Raby, *Speculative Everything: Design, Fiction, and Social Dreaming* (Cambridge, MA: MIT Press: 2013).
- Abigail C. **Durrant**, John Vines, Jayne Wallace, Joyce S.R. Yee, "Research Through Design: Twenty-First Century Makers and Materialities," in *Design Issues* 33, no. 3 (Summer 2017): 3–10. https://doi.org/10.1162/DESI_a_00447.
- Daniel **Fallman**, and Erik Stolterman, "Establishing Criteria of Rigor and Relevance in Interaction Design Research," *Proceedings of Create10 – The Interaction Design Conference* (2010). https://doi.org/10.14236/ewic/CREATE2010.11.
- Elizabeth **Fisher**, and Rebecca Fortnum, eds., *On Not Knowing: How Artists Think* (London, Black Dog Publishing, 2013).
- Mary **Flanagan**, *Critical Play. Radical Game Design* (Cambridge: The MIT Press, 2009).
- Rebecca **Fortnum**, "Creative Accounting: Not Knowing In Talking and Making," in: Elizabeth Fisher, and Rebecca Fortnum, eds., *On Not Knowing: How Artists Think* (London, Black Dog Publishing, 2013): 70.
- Jane **Fulton** Suri, "Informing Our Intuition: Design Research for Radical Innovation," *Rotman Magazine* (Winter 2008): 52–57.
- Fernando **Galdon**, Ashley Hall, and Laura Ferrarello, "Futuring and Trust; A Prospective Approach to Designing Trusted Futures Via a Comparative Study Among Design Future Models," *Proceedings of DCS Conference, Scenarios, Speculation Strategies* (November 2020).
- Bill **Gaver**, and John Bowers, "Annotated Portfolios," *Interactions 19, no.* 4 (2012): 40–49.
- William W. **Gaver**, "What should we expect from research through Design?" in *Proceedings of the 2012 ACM annual conference on Human Factors in Computing Systems (May 2012): 937–946.* https://doi.org/10.1145/2207676.2208538.
- Annie **Gentes**, *In-Discipline of Design: Bridging the Gap Between Humanities and Engineering (*Springer Nature, 2017).
- Anand **Giridharadas**, *Winners Take All: The Elite Charade of Changing the World* (New York: Knopf, 2018).
- Adam **Greenfield**, *Radical Technologies. The Design of Everyday Life* (London: Verso, 2017).

- Walter **Gropius**, *The New Architecture and the Bauhaus* (Cambridge, MA: The MIT Press, 1965).
- John **Heider**, *The Tao of Leadership: Lao Tzu's Tao Te Ching Adapted for the New Age (Atlanta, Georgia: Humanics New Age, 1986).*
- Joel M. **Hektner**, Jennifer A. Schmidt, Mihaly Csikszentmihalyi eds., *Experience Sampling Method: Measuring the Quality of Everyday Life* (Thousand Oaks, CA: Sage Publications, 2006).
- Marjanne van **Helvert**, *The Responsible Object. A History of Design Ideology for the Future* (Amsterdam: Valiz, 2016).
- Niels **Hendriks**, Karin Slegers, and Pieter Duysburgh, "Codesign With People Living With Cognitive or Sensory Impairments: A Case for Method Stories and Uniqueness," *CoDesign* 11 no. 1 (2015): 70–82.
- Garnet **Hertz**, *Disobedient Electronics. Protest*, January 2018. http://www.disobedientelectronics.com/.
- David **Hesmondhalgh**, *The Cultural Industries, Third Edition* (London: Sage Publishing, 2013).
- Justin **Hess**, and Nicholas Fila, "The Development and Growth of Empathy Among Engineering Students," paper presented at 2016 ASEE Annual Conference, and Exposition (New Orleans, LA, 2016).
- Alan **Hevner**, Salvatore March, Jinsoo Park, and Sudha Ram, "Design Science Research in Information Systems, *MIS Quarterly 28, no.* 1 (March 2004): 75–105.
- Robert **Hewison**, *Cultural Capital* (London: Verso, 2014).
- James **Hodge**, Kyle Montague, Sandra Hastings, Kellie Morrissey, "Exploring Media Capture of Meaningful Experiences to Support Families Living with Dementia," in *Proceedings of the 2019 CHI Conference on Human Factors in Computing Systems* (May 2019): 1–14. https://doi.org/10.1145/3290605.3300653.
- Herman van **Hoogdalem**, and Gijs Wanders, G*ezichten van Dementie* (Zwolle: WBOOKS, 2016) en: Herman van Hoogdalem en Constance de Vries, *Mag Ik Gaan. Leven en Sterven met Dementie* (Zwolle: WBOOKS, 2020).
- Maarten **Houben**, Rens Brankaert, Saskia Bakker, Gail Kenning, Inge Bongers, Berry Eggen, "The Role of Everyday Sounds in Advanced Dementia Care," in *Proceedings of the 2020 CHI Conference on Human Factors in Computing Systems* (April 2020): 1–14. https://doi.org/10.1145/3313831.3376577.

- Caroline **Hummels**, and Joep Frens, "Designing for the Unknown: A Design Process for the Future Generation of Highly Interactive Systems and Products, "in *Proceedings of the 10th International Conference on Engineering and Product Design Education* (Barcelona, September 2008): 204–209.
- Sean **Hunt**, Wentao Yuan, Saahil Claypool, and Antonio Ferreira, *Promoting Open Source Models in the Danish Manufacturing Industry* (Worcester, MA: Worcester Politechnic Institute, 12 October 2017),
- Wijnand **Ijsselsteijn**, Ans Tummers-Heemels, Rens Brankaert, "Warm Technology: A Novel Perspective on Design for and with People Living with Dementia," in Rens Brankaert, and G. Kenning eds., *HCI and Design in the Context of Dementia* (Cham: Springer International Publishing, 2020): 33–47. https://doi.org/10.1007/978–3-030–32835–1_3.
- Tim **Ingold**, *Art, Science and the Meaning of Research*, Keynote lecture presented at the symposium Thought Things (Groningen, November 2017).
- Natasha **Iskander**, "Design Thinking Is Fundamentally Conservative and Preserves the Status Quo," *Harvard Business Review* (September 2018).
- François **Jegou**, and Peter Joore, *Food Delivery Solutions* (Cranfield: Cranfield Publishers, 2004).
- Roland **Jochem**, "The Future of Product Creation Is Open and Community-Based," *Research Outreach* 113 (15 April 2020): 6–9. https://doi.org/10.32907/RO-113–69.
- Peter **Jones**, "Systemic Design Principles for Complex Social Systems," *Social Systems and Design* (Springer: Tokyo, 2014): 91–128.
- Peter **Jones**, "Contexts of Co-Creation: Designing with System Stakeholders," *Systemic Design* (Springer: Tokyo, 2018): 3–52.
- Jos De **Jonge**, *Praktijkgericht Onderzoek bij Lectoraten van Hogescholen* (Den Haag: Rathenau Instituut, 2016).
- Peter **Joore**, *New to Improve: The Mutual Influence Between New Products and Societal Change Processes*, (PhD dissertation, Delft University of Technology, 2010).
- Peter **Joore**, and Han Brezet, "A Multilevel Design Model – The Mutual Relationship Between Product-Service System Development and Societal Change Processes," *Journal of Cleaner Production* 97 (2015), 92–105. https://doi.org/10.1016/j.jclepro.2014.06.043.

- Peter **Joore**, Michel van Schie, *Eindrapportage MOVE – Mobiliteitsconcept voor Individueel Transport voor de Korte Afstand – MITKA. TNO Rapport 01/PO/1311/PJO* (Delft: TNO, 2001).
- Gesche **Joost**, Katharina Bredies, Michelle Christensen, Florian Conradi, and Andreas Unteidig eds., *Design as Research: Positions, Arguments, Perspectives* (Basel: Birkhäuser, 2016): 224.
- Tom **Kelley**, *The Art of Innovation: Lessons in Creativity from IDEO, America's Leading Design Firm*, (New York: Crown Publishing Group, 2007).
- René **Kemp**, Johan Schot, and Remco Hoogma, "Regime Shifts to Sustainability Through Processes of Niche Formation: The Approach of Strategic Niche Management," *Technology Analysis & Strategic Management*, 10:2 (1998), 175–198. https://doi.org/10.1080/09537329808524310.
- René **Kemp**, Suzanne van den Bosch, *Transitie-Experimenten – Praktijkexperimenten met de Potentie om bij te dragen aan Transities* (Delft: Kenniscentrum voor Duurzame Systeeminnovaties en Transities, 2006).
- Mahmoud **Keshavarz**, *The Design Politics of the Passport* (London: Bloomsbury, 2019).
- Ruth **Kinna**, and Gillian Whitely, *Cultures of Violence. Visual Arts and Political Violence* (London: Routledge, 2020).
- Lauri **Koskela**, Sami Paavola, Ehud Kroll, "The Role of Abduction in Production of New Ideas in Design," in Pieter E. Vermaas and Stéphane Vial (Eds.), *Advancements in the Philosophy of Design* (Springer International Publishing, 2018): 153–183.
- Ilpo **Koskinen**, John Zimmerman, Thomas Binder, Johan Redström, and Stephan Wensveen, *Design Research Through Practice: From the Lab, Field, and Showroom* (Amsterdam: Elsevier, 2011).
- John **Krogstie**, "Bridging Research and Innovation by Applying Living Labs for Design Science Research," in *Lecture Notes in Business Information Processing* 124 (2012): 161–176. https://doi.org/10.1007/978-3-642-32270-9_10.
- Roman **Krznaric**, *The Good Ancestor. How to Think Long Term in a Short Term World* (London: Penguin Group, 2020).
- Vijay **Kumar**, *101 Design Methods: A Structured Approach for Driving Innovation in Your Company* (Hoboken: John Wiley and Sons, 2012).

- Pepijn de **Lange**, "Nederlanders Eten Van Alle Europeanen de Meeste Vleesvervangers," *De Volkskrant* (10 May 2021).

- Larry **Laudan**, *Progress and Its Problems: Towards a Theory of Scientific Growth* (Berkeley, CA: University of California Press, 1978).

- Amanda **Lazar**, Caroline Edasis, Anne Marie Piper, "A Critical Lens on Dementia and Design in HCI," in *Proceedings of the 2017 CHI Conference on Human Factors in Computing Systems* (Denver: ACM Press, 2017). https://doi.org/10.1145/3025453.3025522.

- Jung-Joo **Lee**, Miia Jaatinen, Anna Salmi, Tuuli Mattelmäki, Riitta Smeds, and Mari Holopainen, "Design Choices Framework for Co-creation Projects," *International Journal of Design* 12, no. 2 (2018): 15–31.

- Jeanne **Liedtka**, "In Defense of Strategy as Design," *California Management Review* 42, no. 3 (2000): 8–30. https://doi.org/10.2307/41166040.

- Jeanne **Liedtka**, "Perspective: Linking Design Thinking with Innovation Outcomes Through Cognitive Bias Reduction," *Journal of Product Innovation Management 32, no. 6 (25 March 2014):* 925–938. https://doi.org/10.1111/jpim.12163.

- Jonas **Löwgren**, "Annotated Portfolios and Other Forms of Intermediate-Level Knowledge," *Interactions*, (February 2013): 30–34.

- Maarit Anna **Mäkelä** and Nithikul Nimkulrat, "Reflection and Documentation in Practice-Led Design Research," *in Proceedings of the 4th Nordic Design Research Conference* (Helsinki, 2011).

- Matt **Malpass**, *Critical Design in Context. History, Theory, and Practice* (London: Bloomsbury, 2017).

- Anu **Manickham**, and Karel van Berkel, *Wicked World: Complex Challenges and Systems Innovations* (Groningen: Noordhoff Uitgevers, 2020).

- Ezio **Manzini**, Luisa Collina, Stephen Evans, *Solution Oriented Partnership. How to Design Industrialised Sustainable Solutions* (Cranfield: Cranfield Publishers, 2004).

- Ezio **Manzini**, *Design, When Everybody Designs: An Introduction to Design for Social Innovation* (Cambridge, MA: MIT Press, 2015).

- Claudia **Mareis**, and Nina Paim, *Design Struggles. Intersecting Histories, Pedagogies, and Perspectives* (Amsterdam: Valiz, 2021).

- Roger L. **Martin**, *The Design of Business: Why Design Thinking Is the Next Competitive Advantage* (Boston, MA: Harvard Business Review Press, 2009).
- Angela **McRobby**, *Be Creative: Making a Living in the New Culture Industries* (Cambridge: Polity Press, 2016).
- Peter **Miller**, "Reliability," in *The SAGE Encyclopedia of Qualitative Research* (Thousand Oaks, CA: SAGE Publications, Inc., 2008): 753–754.
- Peter **Miller**, "Validity," in *The SAGE Encyclopedia of Qualitative Research* (Thousand Oaks, CA: SAGE Publications, Inc., 2008): 909–910.
- Johanneke **Minnema**, Lisa Rosing, Marjolein van Vucht, eds., *Veerkracht – Kennis- en Innovatieagenda voor de Creatieve Industrie 2020–2023* (Eindhoven: CLICKNL, 2020).
- Evgeny **Morozov**, *To Save Everything, Click Here* (New York: PublicAffairs, 2013).
- Manon **Mostert** – Van der Sar, *Hey Teacher, Find Your Inner Designer* (Amsterdam: Boom Uitgevers, 2019).
- Oli **Mould**, *Against Creativity* (London: Verso, 2018).
- Nicoline **Mulder**, *Valuebased Project Management. Een Aanpak voor Chaordische Projecten vanuit het Perspectief van het Complexiteitsdenken* (PhD Thesis, TU Eindhoven, 2012). https://doi.org/10.6100/IR740171.
- Harold G. **Nelson**, and Erik Stolterman, *The Design Way, Second Edition. Intentional Change in an Unpredictable World* (Cambridge, MA: The MIT Press, 2012).
- Safiya Umoja **Noble**, *Algorithms of Oppression* (New York: NYU Press, 2018).
- Don **Norman**, *Living With Complexity* (Cambridge, MA: MIT Press, 2010). Bruce **Nussbaum**, *Is Humanitarian Design the New Imperialism?*, 7 June 2010. https://www.fastcompany.com/1661859/is-humanitarian-design-the-new-imperialism.
- Justin **O'Connor**, "The Great Deflation. Arts and Culture After the Creative Industries," *Making & Breaking* 2 (2021.
- Jenny **Odell**, *How To Do Nothing, Resisting the Attention Economy* (Brooklyn, NY: Melville House, 2019).
- **OECD**, *Frascati Manual 2015: Guidelines for Collecting and Reporting Data on Research and Experimental Development, the Measurement of Scientific, Technological and Innovation Activities* (Paris: OECD Publishing, 2015).
- Sebastian **Olma**, *In Defence of Serendipity. For a Radical Politics of Innovation* (London: Repeater Books, 2016).

- Gerard van **Os** and Karin van Beurden, "Emogram: Help (Student) Design Researchers Understanding User Emotions in Product Design," *Proceedings of the 21st International Conference on Engineering and Product Design Education* (E&PDE 2019) (Glasgow, September 2019). https://doi.org/10.35199/epde2019.44.
- Anja **Overdiek**, and Gary Warnaby, "Co-Creation and Co-Design in Pop-Up Stores: The Intersection of Marketing and Design Research?, *Creativity and Innovation Management 29* (2020): 63–74.
- Anja **Overdiek**, and Heleen Geerts eds., *Innoveren met Labs, Hoe Doe Je Dat? Ervaringen van Future-Proof Retail* (Den Haag: De Haagse Hogeschool, 2020).
- Victor **Papanek**, *Design for the Real World. Human Ecology and Social Change* (St Albans : Paladin, 1974).
- Victor **Papanek**, and James Hennessey, *How Things Don't Work* (New York: Pantheon Books, 1977).
- Dominic **Pettman**, *Infinite Distraction* (Cambridge: Polity Press, 2016).
- **Hasso Plattner** Institute of Design, *An Introduction to Design Thinking: Process Guide* (Stanford: 2010).
- Michael **Polanyi**, *The Tacit Dimension* (New York: Doubleday Anchor, 1966).
- Rebecca Anne **Price**, Christine De Lille and Katinka Bergema, "Advancing Industry Through Design: A Longitudinal Case Study of the Aviation Industry," *She Ji: The Journal of Design, Economics, and Innovation* 5, no. 4 (2019): 304–326.
- Kate **Raworth**, *Doughnut Economics, Seven Ways to Think Like a 21st-Century Economist* (White River Junction, VT: Chelsea Green Publishing, 2017).
- Elizabeth **Resnick**, *The Social Design Reader* (London: Bloomsbury, 2019).
- Holger **Rhinow**, Eva Köppen, Christoph Meinel, "Design Prototypes as Boundary Objects in Innovation Processes," in *Proceedings of the Design Research Society International Conference* (Bangkok, July 2012): 1581–1590.
- Linda **Rindertsma** ed., *Kennis- en Innovatieagenda voor de creatieve industrie 2020–2023* (Eindhoven: TKI CLICKNL, 2020).
- Horst W.J. **Rittel** and Melvin M. Webber, "Dilemmas in a General Theory of Planning," *Policy sciences*, 4, no. 2 (1973): 155–169.
- Ken **Robinson**, *Out of Our Minds, The Power of Being Creative* (Hoboken NJ: Wiley, 2011).

- Yvonne **Rogers**, "New Theoretical Approaches for HCI," *Annual Review of Information, Science and Technology* 38 (2004): 87–143.
- Donald **Ropes**, *Organizing Professional Communities of Practice* (Amsterdam: University of Amsterdam Press, 2010).
- Hartmut **Rosa**, *Resonanz: Eine Soziologie der Weltbeziehung* (Frankfurt am Main: Suhrkamp, 2016).
- Hartmut **Rosa**, *Unverfügbarkeit (Unruhe bewahren)* (Salzburg: Residenz. 2018).
- Daniela K. **Rosner**, *Critical Fabulations. Reworking the Methods and Margins of Design* (Cambridge: The MIT Press, 2018).
- George **Roth** and Art Kleiner, *Field Manual for a Learning Historian* (Boston: MIT, 1996).
- Michael **Rubenstein**, Alejandro Cornejo, Radhika Nagpal, "Programmable Self-Assembly in a Thousand-Robot Swarm," Science 345, no. 6198 (15 Aug 2014*):* 795–799. *https*://doi.org/10.1126/science.1254295.
- Elizabeth **Sanders** and Pieter Jan Stappers, *Convivial Toolbox. Generative Research for the Front End of Design* (Amsterdam: BIS publishers, 2012).
- Elizabeth **Sanders** and Pieter Jan Stappers, "Co-creation and the New Landscapes of Design," *CoDesign* 4 no. 1 (2008): 5–18. https://doi.org/10.1080/15710880701875068.
- Daniela **Sangiorgi**, "Transformative Services and Transformation Design," *International Journal of Design* 5, no. 2 (2011).
- Donald **Schön**, *The Reflective Practitioner: How Professionals Think in Action* (New York, Basic Books, 1984).
- Tristan **Schultz**, Danah Abdulla, Ahmed Ansari, Ece Canlı, Mahmoud Keshavarz, Matthew Kiem, Luiza Prado de O. Martins & Pedro J.S. Vieira de Oliveira (2018) Editors' Introduction, *Design and Culture,* 10:1, 1–6, DOI: 10.1080/17547075.2018.1434367.
- Joseph **Schumpeter**, *The Theory of Economic Development* (Cambridge: Harvard University Press, 1911).
- Maung **Sein**, Ola Henfridsson, Sandeep Purao, Matti Rossi and Rikard Lindgren, "Action Design Research," *MIS Quarterly* 35 (2011), 37–56. https://doi.org/10.2307/23043488.
- Tim **Seitz**, *Design Thinking and the New Spirit of Capitalism. Sociological Reflections on Innovation Culture* (Cham: Palgrave Pivot, 2020).

- Frans **Sengers**, Anna J. Wieczorek, Rob Raven, "Experimenting for Sustainability Transitions: A Systematic Literature Review," *Technological Forecasting and Social Change* 145 (2019), 153–164.
- Birger **Sevaldson** at www.systemsoriented-design.net, visited January 2021.
- Herbert **Simon**, *The Sciences of the Artificial*, Third Edition (Cambridge, MA: MIT Press, 1996).
- Jesper **Simonsen** and Toni Robertson eds., *Routledge International Handbook of Participatory Design* (Oxfordshire: Routledge, 2012).
- Wina **Smeenk**, *Navigating Empathy: Empathic Formation in Co-Design*, PhD Thesis, TU Eindhoven (2 December 2019).
- Wina **Smeenk**, Anja Köppchen, and Gène Bertrand, *Het Co-Design Canvas. Een Empatisch Co-Design Instrument met Maatschappelijke Impact* (InHolland, 2020).
- Wina **Smeenk**, Janienke Sturm, and Berry Eggen, "Empathic Handover: How Would You Feel? Handing Over Dementia Experiences and Feelings in Empathic Co-Design," *International Journal of CoCreation in Design and the Arts* 14 no. 4 (2018). https://doi.org/10.1080/15710882.2017.1301960.
- Wina **Smeenk**, Janienke Sturm, and Berry Eggen, "Comparison of Existing Frameworks Leading to an Empathic Formation Compass for Co-design," *International Journal of Design* 13, no. 3 (2019: 53–68.
- Wina **Smeenk**, Oscar Tomico and Koen van Turnhout, "A Systematic Analysis of Mixed Perspectives in Empathic Design: Not One Perspective Encompasses All," *International Journal of Design* 10, no. 2 (2016).
- Aletta **Smits**, Erik Hekman, Koen van Turnhout, "Ear to the Ground: Using Text Mining to Pick Up All Sudanese Voices for Radio Dabanga," *The EuroIA Conference* (Kraków, September 2020).
- Marie L. J. **Søndergaard**, *"Staying With the Trouble Through Design: Critical-feminist Design of Intimate Technology,"* PhD Thesis (Aarhus University, 3 December 2018).
- Nick **Srnicek**, *Platform Capitalism* (Cambridge: Polity Press, 2016).
- Pieter Jan **Stappers** and Elisa Giaccardi, "Research Through Design," in *The Encyclopedia of Human-Computer Interaction, 2nd edition, eds.* Mads Soegaard, and Rikke Friis-Dam (Aarhus, Denmark: 2017): 1–94.
- Guido **Stompff**, *Design Thinking- Radicaal Veranderen in Kleine Stappen* (Amsterdam: Boom uitgevers, 2018)

- Guido **Stompff**, *De Kracht van Verbeelden, Design Thinking in Teams,* Inaugurele Rede (Amsterdam: Hogeschool Inholland, 2020).
- Lucy **Suchman**, "Working Relations of Technology Production and Use," *Computer Supported Cooperative Work 2*, no. 1 (1994): 21–39.
- Laurent de **Sutter**, *Narcocapitalism* (Cambridge: Polity Press, 2018).
- Halina **Szejnwald** Brown, Philip Vergragt, Ken Green, Luca Berchicci, "Learning for Sustainability Transition Through Bounded Socio-technical Experiments in Personal Mobility," *Technology Analysis & Strategic Management*, 15:3 (2003), 291–315. https://doi.org/10.1080/09537320310001601496.
- Bruce M. **Tharp**, and Stephanie M.Tharp, *Discursive Design. Critical, Speculative, and Alternative Things* (Cambridge: The MIT Press, 2019).
- Julie **Thompson** Klein, "Prospects for Transdisciplinarity," *Futures* 36 no. 4 (2004): 515–526. https://doi.org/10.1016/j.futures.2003.10.007.
- Nato **Thompson** & Gregory Sholette, *The Interventionists, Users' Manual for the Creative Disruption of Everyday Life* (Cambridge: MIT Press, 2004).
- Ferdinand **Tönnies**, *Gemeinschaft und Gesellschaft (1887)* (Whitefish, MT: Literary Licensing, 2014).
- Peter **Troxler**, "The Beginning of a Beginning of the Beginning of a Trend," in Bas van Abel, Roel Klaassen, Lucas Evers, Peter Troxler, *Open Design Now: Why Design Cannot Remain Exclusive* (Amsterdam: BIS Publishers, 2011).
- Peter **Troxler**, "Building Open Design as a Commons" in Loes Bogers and Letizia Chiappini, *The Critical Makers Reader: (Un)Learning Technology* (Amsterdam: Institute of Network Cultures, 2019): 2018–226.
- Peter **Troxler**, Eva Visser, and Maarten Hennekes, *Roadmap Makerplaatsen. Van Knutselen 2.0 Naar Leren met 21ste Eeuwse Vaardigheden*, (Rotterdam: Kenniscentrum Creating 010, 2018).
- Peter **Troxler**, and Patricia Wolf, "Digital Maker-Entrepreneurs in Open Design: What Activities Make Up Their Business Model?," *Business Horizons* 60, no. 6 (1 November 2017): 807–17. https://doi.org/10.1016/j.bushor.2017.07.006.

- Peter **Troxler** and Patricia Wolf, "Look Who's Acting! Applying Actor Network Theory For Studying Knowledge Sharing in a Co-Design Project, *International Journal of Actor-Network Theory and Technological Innovation 7*, no. 3 (2015).
- Koen van **Turnhout**, Arthur Bennis, Sabine Craenmehr, Robert Holwerda et al, "Design Patterns for Mixed-Method Research in HCI," *Proceedings of the 8th Nordic Conference on Human-Computer Interaction: Fun, Fast, Foundational* (October 2014): 361–370.
- Koen van **Turnhout**, Sabine Craenmehr, Robert Holwerda, Mike Menijn, Jan-Pieter Zwart, René Bakker, "Tradeoffs in Design Research: Development Oriented Triangulation," in *Proceedings of the 27th International BCS Human Computer Interaction Conference* (September 2013).
- Koen van **Turnhout**, Marjolein Jacobs, Miriam Losse, Thea van der Geest, René Ronald Bakker, "A Practical Take on Theory in HCI," *White paper* (2019).
- Koen van **Turnhout**, and Aletta Smits, "On Solution Repertoire," in: *Proceedings of the 23rd Engineering and Product Design Education Conference (*Herning, Denmark, 2021).
- Koen van **Turnhout**, Stijn Hoppenbrouwers, Paul Jacobs, Jasper Jeurens, Wina Smeenk, and René Ronald Bakker, "Requirements From The Void: Experiences With 1: 10: 100," in *Proceedings of the 3rd Workshop on Creativity in Requirements Engineering* (Essen, 2011).
- Merlijn **Twaalfhoven**, *Het Is Aan Ons. Waarom We de Kunstenaar in Onszelf Nodig Hebben om de Wereld te Redden* (Amsterdam/Antwerpen: Atlas Contact, 2020).
- Job van 't **Veer**, Eveline Wouters, Remko van der Lugt, Monica Veeger, *Ontwerpen Voor Zorg en Welzijn* (Bussum: Coutinho, 2020).
- Roberto **Verganti**, *Design-Driven Innovation, Changing the Rules of Competition by Radically Innovating What Things Mean* (Boston, MA: Harvard Business Press Books, 2009).
- **Verheijen**, L., Praasterink, P. Giezen, P, van Aken, S., and Riedesel, A., *Student Manual: A Food Systems Approach: A Toolkit to Unravel Complexity* (Research Group Future Food Systems, HAS University of Applied sciences, 2020).
- Theresa **Willingham** and Jeroen De Boer, *Maker-spaces in Libraries, Library Technology Essentials 4* (Lanham MD: Rowman & Littlefield Publishers, 2015).
- Ludwig **Wittgenstein**, *Tractatus Logico-Philosophicus* (Londen: Kegan Paul, Trench, Trubner & Co, 1922).

- Patricia **Wolf**, and Peter Troxler, "Community-Based Business Models: Insights From an Emerging Maker Economy," *Interaction Design and Architecture(s)* 30 (2016): 75–94.

- Ernst von **Weizsäcker**, Amory B. Lovins, L. Hunter Lovins, *Factor Four: Doubling Wealth, Halving Resource Use* (London: Earthscan Publications Ltd, 1998).

- Eveline **Wouters**, and Joost van Hoof, "Professionals' Views of the Sense of Home in Nursing Homes: Findings From LEGO SERIOUS PLAY Workshops" *Gerontechnology* 16 (2017): 218–223. https://doi.org/10.4017/gt.2017.16.4.003.00.

- Eveline **Wouters**, and Sil Aarts, *Ethiek van Praktijkgericht Onderzoek: Zonder Ethiek is het al Moeilijk Genoeg.* (Houten: Bohn Stafleu van Loghum, 2017).

- Marieke **Zielhuis**, Froukje Sleeswijk Visser, Daan Andriessen, Pieter Jan Stappers, "What Makes Design Research More Useful for Design Professionals? An Exploration of the Research-Practice Gap," *Journal of Design Research* (in press).

- John **Zimmerman**, Jodi Forlizzi, and Shelley Evenson, "Research Through Design as a Method for Interaction Design Research in HCI," in *Proceedings of the SIGCHI Conference on Human Factors in Computing Systems* (2007): 493–502. https://doi.org/10.1145/1240624.1240704.

- John **Zimmerman**, Erik Stolterman, and Jodi Forlizzi, "An Analysis and Critique of Research Through Design: Towards a Formalization of a Research Approach," in *Proceedings of the 8th ACM Conference on Designing Interactive Systems – DIS'10* (2010): 310–319.

- Slavoj **Zizek**, *The Courage of Hopelessness, Chronicles of a Year of Acting Dangerously,* (London: Penguin Books, 2017).

- Tonnie van der **Zouwen**, *Actieonderzoek Doen: Een Routewijzer voor Studenten en Professionals* (Amsterdam: Boom uitgevers, 2018).

- Antien **Zuidberg**, *What U Design = How U Design*, Inaugurele Rede ('s Hertogenbosch: HAS Hogeschool, 2020)